CROP PRODUCTION SCIENCE IN HORTICULTURE

Series Editors: Jeff Atherton, Senior Lecturer in Horticulture, University of Nottingham, and Rees, Horticultural Consultant and Editor, *Journal of Horticultural Science.*

This series examines economically important horticultural crops selected from the major production systems in temperate, subtropical and tropical climatic areas. Systems represented range from open field and plantation sites to protected plastic and glass houses, growing rooms and laboratories. Emphasis is placed on the scientific principles underlying crop production practices rather than on providing empirical recipes for uncritical acceptance. Scientific understanding provides the key to both reasoned choice of practice and the solution of future problems.

Students and staff throughout the world involved in courses in horticulture, plant science, food science and applied biology level will welcome this series as a succinct and source of information. The books will also be invaluable to progressive advisers and product users requiring an authoritative, but brief scientific introduction to particular crops or systems. Keen gardeners wishing to understand the scientific of recommended practices will also find the series useful.

The authors are internationally recognized experts with extensive experience of their subjects in a common format covering all aspects of production, from breeding, to propagation and planting, through to harvesting, handling and storage. Selective references are included to direct the reader to further information on specific topics.

Titles Available:
1. **Ornamental Bulbs, Corms and Tubers** A.R. Rees
2. **Citrus** F.S. Davies and L.G. Albrigo
3. **Onions and Other Vegetable Alliums** J.L. Brewster
4. **Ornamental Bedding Plants** A.M. Armitage
5. **Bananas and Plantains** J.C. Robinson
6. **Cucurbits** R.W. Robinson and D.S. Decker-Walters
7. **Tropical Fruits** H.Y. Nakasone and R.E. Paull

LETTUCE, ENDIVE AND CHICORY

E.J. Ryder
US Department of Agriculture
Agricultural Research Service
Salinas
California
USA

CABI *Publishing*

CABI *Publishing* – a division of CAB INTERNATIONAL

CABI *Publishing*
CAB INTERNATIONAL
Wallingford
Oxon OX10 8DE
UK

CABI *Publishing*
10 E 40th Street,
Suite 3203
New York, NY 10016
USA

Tel: +44 (0)1491 832111
Fax: +44 (0) 1491 833508
Email: cabi@cabi.org

Tel: +1 212 481 7018
Fax: +1 212 686 7993
Email: cabi-nao@cabi.org

A catalogue record for this book is available from the British Library, London, UK.

Library of Congress Cataloging-in-Publication Data
Ryder, Edward J., 1929-
    Lettuce, endive, and chicory / E.J. Ryder.
        p.   cm. — (Crop production science in horticulture; 7)
    Includes bibliographical references (p.   ) and index.
    ISBN 0-85199-285-4 (alk. paper)
        1. Lettuce. 2. Endive. 3. Chicory. I. Title. II. Series.
    SB351.L6R94   1998
    635´.5—dc21

98-34438
CIP

ISBN 0 85199 285 4

Typeset in Britain by Solidus (Bristol) Ltd
Printed and bound in the UK at the University Press, Cambridge

CONTENTS

PREFACE

In writing this book, I have kept in mind not only the goals of the Crop Production Science in Horticulture series, but also a few of my own. Accordingly, I hope the book provides a useful overall view of the scientific bases for the production and marketing of the three salad vegetables discussed. In addition, I wish the book to be the entry point for more extensive acquaintance with the sciences. Therefore, I have been very liberal in citing references to the original works discussed briefly here, in order to provide the hungrier reader with the opportunity to explore the literature on the genetics, breeding, physiology, pathology, entomology and economics of these crops. Citations in the three production chapters also afford the opportunity to search the literature for more information on growing, harvesting, marketing and seed production of the crops. I have also called attention to several books and other publications that will provide fuller discussions of some topics than those included here.

The reader will note at least two apparent disparities in scope of the discussions. First, by far the most extensive discussions are on lettuce. Lettuce is much more widely grown than either endive or chicory, and the amount of literature available on lettuce is considerably greater. The combination of economic importance and published research accounts for the difference in coverage of lettuce as compared with endive and chicory. However, as noted in several places in the text, many of the principles and practices attributed to lettuce also apply to endive and chicory. Second, lettuce is more extensively grown in the USA than in any other country. Consequently, much of the published literature and production information available on that crop comes from the USA. I have tried to cover lettuce research and production in other countries as thoroughly as I could in order to partially redress that disparity. The reader can judge how successful I have been. Certainly, the principles explored and information developed are often universally applicable wherever they were first described.

The transfer of information around the world is getting easier each year.

Publication of research results has long been available worldwide, although with some difficulties in some parts of the world. As the electronic revolution continues to develop, more information will become easily available in more places. In addition, those doing research, extension work and production of leafy vegetables have the opportunity to participate in two international forums to discuss aspects of work in these crops. The Eucarpia Meeting on Leafy Vegetables has occurred periodically since 1976 in European cities. The International Lettuce Conference and Leafy Vegetables Workshop has been held every two or three years in various production areas of the USA since 1965. These two conference series offer the opportunity for face-to-face conversations with other practitioners in the field, as well as the dissemination of information in a more formal manner.

The length of Chapters 2 and 7 reflects my own long activity as lettuce breeder and geneticist, and the emphasis in my breeding programme on breeding for disease resistance. I recently completed 41 years in the position I hold. This length of service gives one some perspective over time, which undoubtedly accounts for the extensive discussions on the history of lettuce and of lettuce breeding. I hope these will prove of some interest to the reader.

I have asked a large number of people for help in writing this book and I take this occasion to thank them. For providing information on the crops and their production, I am indebted to: George Emery, Maria L. Gómez-Guillemón, Kees Reinink, Herve Michel, David A.C. Pink, Ron van der Laan, Gadi Tsafrir, Zvi Karchi, Ryoichi 'Joe' Kojima, Dan Trimboli, Leigh James, C. Beckingham, Bernard Moreau, David C.E. Wurr and James E. Manassero. For reviewing chapters or sections of the book, I thank: Kent J. Bradford, Richard W. Michelmore, William Waycott, Vincent E. Rubatzky, James E. Duffus, Krishna V. Subbarao, Carolee Bull, Mohammad A. Bari, Steven A. Fennimore, James E. Manassero and Edward A. Kurtz.

Some of the information reported here is from my own research as lettuce breeder for the US Department of Agriculture–Agricultural Research Service. Much of the research was generously supported by the California Lettuce Advisory Board. I am grateful for that support and for the guidance provided by the Board managers, Edward A. Kurtz and the late Harold G. Bradshaw.

I wish to thank the publisher, CABI *Publishing*, for two reasons. One is for the excellent guidance and advice provided, especially by Tim Hardwick and also by Martin Townsend and Ali West, and the fine useful reviews by A.R. Rees, J. Atherton and the anonymous reviewer. The other is for a generous contribution to the Endowment Fund of the American Society for Horticultural Science in lieu of royalties for the author. Finally, I give special thanks to my wife, Elouise, for her heroic work on the computer, helping to compile the index, not to mention her patience, love and fortitude while the book was being written.

INTRODUCTION TO THE CROPS

> The queen of the salad plants, the leaf that is synonymous with salad in most of the temperate zone, that exquisite, crisp foundation for the salad dish that also shreds into a fine and almost neutral medium for a variety of tasty sauces, is, of course, lettuce, *Lactuca sativa* L.

This sentence introduces the lettuce chapter in the book *Edible Leaves of the Tropics* (Martin and Ruberte, 1975). Many different kinds of leaves, as well as other plant parts, appear in salads over the world, but none are as ubiquitous as lettuce. To a lesser extent, endive and chicory, sometimes incorrectly called lettuces, also contribute to the cool part of the meal, the salad.

Lettuce is grown commercially in many countries in the world, particularly in North America, western Europe, the Mediterranean basin, Australia and parts of Asia. It is also grown in home gardens in most of these countries. Commercial growers emphasize the production of lettuces that form heads, while home gardeners are more likely to plant leaf or cutting lettuces, which are easier to grow and harvest.

Endive and chicory are most commonly found in western Europe, although their usage is increasing in North America. Endive is grown primarily for raw use in salads, while chicory in both the forced and unforced forms is often cooked as well.

Lettuce is a staple crop, especially in the USA. The term staple is used in the sense that it is consumed regularly, by large numbers of people, across ethnic and class lines. In this sense, it is in the same class as potatoes, tomatoes, cabbage, onions and beans.

USES

Lettuce, endive and chicory leaves are used overwhelmingly as a raw product in salads. Occasionally they are lightly cooked or put in soups.

Until recently, consumers in most countries preferred one type of lettuce over others that might be available. In the USA, for example, crisphead lettuce, known more familiarly as iceberg lettuce, made up over 90% of the total commercial crop. In the UK and Northern Europe, the butterhead type was the overwhelming favourite, while in the Mediterranean basin, cos, or romaine, types were preferred. In recent years, this picture has changed. As much as 25–30% of the crop in the USA is now romaine, butterhead and leaf lettuce. In western Europe and other parts of the world, the crisphead type has become popular, and now comprises the major lettuce group in the UK and in the Scandinavian countries.

The steadily growing uses of salad leaves include a variety of value-added products. For example, crisphead lettuce is harvested in bulk bins, cored, chopped or shredded, washed and packed in various-sized plastic bags. These are shipped in containers to fast-food chains and other restaurants, and to institutions such as hospitals and schools. The lettuce may be premixed with other salad vegetables, such as carrots and red cabbage, or mixed at the receiving point.

These cut materials may also be put into a plastic package sufficient for one or two meals. Consumer packages are rapidly increasing in popularity. They consist of a wide variety of mixtures of lettuces, endive, chicory, spinach and other vegetables. They may also include small packages of dressing, croutons, grated cheese or other peripheral items.

Another type of value-added product is mesclun, a mixture of baby leaves. The mixture includes several different lettuces, chicory, endive and other leaf forms unrelated to the composite types. The leaves are picked while young and small and the mixtures are sold in bulk in the market.

The chicon is the value-added product of forced chicory regrowth. The chicory is field-grown as a row crop. When mature, the green tops are removed and the roots replanted in an enclosed area so that the second growth of the top takes place in the dark. A tight pointed head is formed, which is white or pale yellow. This is the chicon and it can be eaten in a salad or as a cooked vegetable.

There are a number of other food and non-food uses for these leafy species. A primitive form of lettuce, grown in Egypt, and probably elsewhere in the Mediterranean area, has large seeds. The seeds are pressed to express an oil that is used for cooking.

Stem lettuce, as the name indicates, forms a thick elongated stem, which can be eaten raw or cooked. Raw use is common in Egypt; stem lettuce is used as a cooked vegetable in China.

A wild lettuce, *Lactuca brevirostris* Benth., from South America, produces large leaves that become brown, relatively tough and thick when dried. These are used as a tobacco substitute from which nicotine-free cigarettes are manufactured.

The milky juice or latex of several *Lactuca* species contains two sesquiterpene lactones called lactucin and lactucopicrin. These have a sedative or narcotic effect; extracts from *Lactuca virosa* especially are used as sleep inducers and cough suppressants in Europe. Lettuce-seed oil has also shown analgesic and sedative properties.

The pressing of lettuce seed for oil is an ancient practice in Egypt, which may antedate the use of lettuce as a vegetable. There is still some production of lettuce seed for oil in Egypt. The oil can be extracted by various types of presses. It can also be extracted with certain solvents, such as acetone and ether (Ramadan, 1976). The two principal fatty acids present in the oil are linoleic and oleic acids (Shoeb *et al.*, 1969).

Chicory roots can be processed for two different types of use. The roots of one group of cultivars are ground up and added to coffee to modify the flavour or are used alone as a substitute for coffee. During the Second World War, when coffee was scarce on the home front, this use was more common than at present.

Chicory roots have a high content of inulin, which can be converted industrially to its constituent fructose sugars. Fructose sugar is used as a sweetening component in sweets, syrup and soft drinks.

Recently, chicory has become useful as a forage crop. The cultivar Grasslands Puna was developed in New Zealand (Moloney and Milne, 1993). It is planted on 8000–10,000 ha annually, usually with clover or grass and clover.

PRODUCTION AND VALUE

Lettuce

Lettuce, in one or more of its various forms, is produced in most temperate and subtropical areas of the world. The time and place of production depend on the hemisphere (north or south) in which production takes place and, within hemispheres, vary with climate and season. Lettuce is most easily produced under relatively cool and mild temperatures, and this requirement dictates the time period in which it is grown in each region.

By far the greatest commercial production of lettuce takes place in the USA, where the salad has become a staple part of most midday and evening meals. Almost as common is the sandwich, served mostly at midday, which usually includes lettuce. In the USA in 1997, lettuce was grown on 113,000 ha, which produced over 3.9 million metric tonnes of product, worth over $1.6 billion ($10^9$) (USDA, 1998) (Tables 1.1 and 1.2). Production is not uniformly distributed over the country, but is concentrated mostly in two states, California and Arizona, which together account for over 90% of the production. California production varies from 70 to 75%, while Arizona contributes about 20%. Five other states, Colorado, New Mexico, New York, New Jersey and Washington, produce nearly all the rest. Texas, Michigan, Florida and Ohio produce smaller amounts. About 75% of American

Table 1.1. Area, production and value of crisphead lettuce in the USA, 1997. Total and by leading states. (From USDA, 1998).

	Area (ha)	Production (million tonnes)	Value (million $)
USA	82,150	3,116	1,188
California	57,090	2,243	947
Arizona	21,900	765	202
Colorado	890	30	10
New Mexico	970	27	13
New Jersey	530	17	10
Washington	410	8	3
New York	320	9	3

Table 1.2. Area, production and value of leaf, butterhead and cos lettuce in the USA, 1997. Total and by leading states. (From USDA, 1998).

	Area (ha)	Production (million tonnes)	Value (million $)
Leaf and butterhead			
USA	17,210	420	245
California	14,450	340	190
Arizona	2,350	71	50
Ohio	260	6	3
Florida	140	2	2
Cos			
USA	13,660	422	171
California	9,720	290	113
Arizona	3,360	120	51
Florida	360	9	4
Ohio	220	4	3

production is of the iceberg type; nearly all the rest is in romaine, leaf and butterhead types.

California produces more lettuce than any country in the world. In 1997, California farmers grew lettuce on 81,300 ha, harvesting nearly 2.9 million tonnes, worth over $1.25 billion. Arizona also exceeded most other countries, producing 960,000 tonnes of lettuce on 28,000 ha, valued at over $300 million.

The principal production areas in California are the Salinas Valley, the Santa Maria Valley, the Imperial Valley and the west side of the San Joaquin Valley. Most production in Arizona is in the south-west corner, near Yuma.

There is substantial production of lettuce in several other countries, particularly in western Europe, where Great Britain, France, Spain, the

Netherlands, Italy, Germany and Belgium produce lettuce (Eurostat, 1998) (Table 1.3). Australia, Japan and Israel are also important lettuce producers. Lesser amounts are produced in several countries in South and Central America, Africa and China.

In the UK, approximately 75% of the production is crisphead lettuce, 15% is butterhead and 10% is cos. This represents an enormous change from the 1970s and earlier, when 80–90% of the production was of the butterhead type. Outdoor lettuce is produced mainly in Kent, Lincolnshire and the Thames Valley. The protected crop in calendar year 1993 occupied 1200 ha. Over half the crop was produced in north-west England, North Wales and north-east England, and the rest in East Anglia and in the south-east.

Production in France includes mostly butterhead and Batavia lettuce for home consumption and iceberg-type crisphead for export. The most important production areas are Provence, Languedoc-Roussillon, Île de France and the south-west. Brittany specializes in the iceberg subtype.

In Germany, about two-thirds of production is butterhead lettuce and one-third crisphead. Most production is in three areas, Baden-Wurtemberg, Rhineland-Westphalia and the Palatinate.

A substantial portion of lettuce production in the Netherlands is in

Table 1.3. Leafy vegetable statistics for leading European Community nations, 1996 (From Eurostat, 1998).

	Area (ha)	Production (1000 tonnes)
Lettuce (all types)		
Spain	33,600	924.6
Italy	21,300	419.7
France	13,500	365.6
UK	7,500	230.5
Germany	5,900	143.8
Greece	3,600	69.5
Belgium	2,500	85.1
Netherlands	2,300	110.0
Endive (narrow and broad leaf)		
Italy	12,700	235.5
France	5,800	139.2
Greece	2,400	40.6
Spain	2,200	50.2
Netherlands	670	39.8
Chicory (witloof and green leaf)		
France	14,300	243.7
Italy	15,900	226.8
Belgium	6,100	78.2
Netherlands	4,300	85.0

greenhouses, and consists primarily of the butterhead type. The outdoor crop consists of butterhead, crisphead and leaf lettuces.

In Belgium, most lettuce is butterhead, and about two-thirds is grown under cover and one-third outdoors.

Lettuce is grown in Spain both for domestic consumption and for export. Most is raised in Murcia, which grows lettuce all year round and produces about one-third of the crop, followed by Andalusia, Valencia and Catalonia. Spain has become an important grower and exporter of iceberg-type lettuce. In 1994, Murcia produced 175,000 tonnes for export on 10,500 ha. Cos and Latin lettuces are also important in Spain, the former primarily for domestic consumption.

Italy is the largest producer of vegetables in western Europe. More than half the lettuce is butterhead lettuce, about one-third is cos, with lesser amounts of crisphead and leaf lettuces.

In Australia, most of the production is centred around large cities, principally in Victoria, Queensland, New South Wales and Western Australia (D. Trimboli, personal communication). Lettuce was produced on about 4000 ha in 1992/93, yielding 99,000 tonnes, worth $43.3 million. Most of the lettuce is of the iceberg type, with lesser amounts of cos and leaf types.

In Japan, iceberg, leaf and butterhead lettuces are grown on 23,000 ha, which produce 520,000 tonnes, valued at $1.5 billion (R. Kojima, personal communication). The main prefectures for growing lettuce are Nagano, Ibaragi, Kagawa and Hyogo. Lettuce is grown all year.

Israel grows lettuce on 1200–1450 ha, about two-thirds of which is in cos lettuce and the remainder in iceberg (Z. Karchi, G. Tsafrir, personal communications). Production is 68,000–90,000 tonnes, which is worth $28.8–34.8 million.

Endive

Endive, including narrow-leaved and broad-leaved types, is grown in relatively small amounts in several countries, mostly in Europe (Eurostat, 1998) (Table 1.3) and North America. The principal producers are Italy, France, Spain, the USA and the Netherlands. Production in the USA is mostly in Florida, New Jersey and Ohio, totalling about 1300 ha and 25,000 tonnes in 1997 (USDA, 1998).

Chicory

Most of the witloof chicory is grown in three countries: France, Belgium and the Netherlands (Eurostat, 1998) (Table 1.3). Most French witloof is produced in Pas de Calais and in Picardy. Belgium grows witloof, mostly in the Brabant and in Flanders. The Netherlands also grows mostly witloof. The most important production areas are North Holland, South Holland, North Brabant and Zeeland. Non-forced types are grown primarily in Italy and France. Italy grows nearly all non-forced chicory. The radicchio types are grown mostly in the north, around Verona, Treviso and Venice. Catalonia types are grown in the south.

NUTRITIONAL VALUE

Lettuce, endive and chicory contribute primarily vitamins and minerals to the human diet, plus substantial fibre and a great deal of water. The nutritional content varies with the degree of leaf colour, green outer leaves having more value than whitish inner leaves. This comparison is also true for leaf lettuce versus iceberg type and green chicory versus the witloof type.

Nutritional values are given for four types of lettuce: crisphead, butterhead, cos and leaf (Rubatzky and Yamaguchi, 1997) (Table 1.4). The cos and leaf types generally exceed the butter and crisp types, especially the latter, in content of the major vitamins and minerals and in fibre content.

The actual nutritional value of lettuce can be related to consumption. Stevens (1974) calculated, for 39 major fruits and vegetables, the nutrient composition based on ten vitamins and minerals. The foods were ranked for each nutrient and the sum of the ranks provided an overall ranking. This procedure was repeated, with the nutrient values weighted by usage of each crop, estimated from the total production in the USA in the year 1970. In the composition-only ranking, lettuce was 26th in nutritional value. In the

Table 1.4. Nutritional values for leafy vegetables. Values are for 100 g of edible product. (From Rubatzky and Yamaguchi, 1997, as compiled from several original sources.)

	Minerals (g)					Vitamins		Water	Fibre
	Ca	P	Fe	Na	K	A(IU)	C(g)	%	g
Lettuce									
Crisp	22	26	1.5	7	166	470	7	95.5	0.5
Butter	35	26	1.8	7	260	1065	8	95.1	0.5
Cos	44	35	1.3	9	277	1925	22	94.9	0.7
Leaf	68	25	1.4	9	264	1900	18	94.0	0.7
Endive	66	41	1.3	50	304	2140	8	93.8	0.9
Chicory									
Witloof	16	20	0.5	8	177	Tr	—	95.1	—
Green	93	43	0.9	45	420	4000	24	92.0	0.8

Ca, calcium; P, phosphorus; Fe, iron; Na, sodium; K, potassium; Tr, trace.

ranking weighted by consumption, lettuce moved up to fourth on the list, which reflects its importance as a major, staple crop. This comparison was for crisphead lettuce specifically. As the proportion of consumption of the other types has increased in recent years, the total contribution of lettuce to the diet has also increased.

The nutritional contribution of endive compares favourably with lettuce, particularly for phosphorus, sodium, potassium, vitamin A and fibre (Table 1.4). Green chicory also compares favourably with lettuce and with endive, while the contribution from witloof has obviously been reduced considerably by the forcing process, which prevents chlorophyll development.

Additional nutritive and other health benefits may come from various biologically active compounds, such as chlorogenic acid, which may contribute anticarcinogenic properties. Research on the value of these compounds is receiving increasing attention.

BOTANY AND TAXONOMY

Lettuce, endive and chicory all belong to the largest dicotyledonous family in the plant kingdom, the *Asteraceae* (*Compositae*). All three are also in the subfamily *Cichorioideae* and the tribe *Lactuceae*. Lettuce is in the genus *Lactuca*, while endive and chicory are in *Cichorium*.

The lettuce plant forms a tap root, relatively thick at the crown and gradually tapering towards the tip, which can grow 60 cm or longer. Lateral roots are formed along the length of the tap root, most densely near the surface and decreasing with increasing depth. Consequently, nutrient and water absorption occurs mostly in the upper levels of the soil (Jackson, 1995).

Lettuce leaves are spirally arranged on a shortened stem, forming a rosette of leaves. Rosette development may continue for the vegetative life of the plant, as in leaf lettuces, or form a rounded head, as in crisphead and butterhead types, or an elongated head, as in the cos type. The earliest leaves are elongated, gradually becoming broader with subsequent leaf formation. The later mature leaves of butterhead, crisphead and some leaf lettuces are broader than long. Leaves of cos lettuce and some leaf lettuces remain longer than broad.

The degree of green colour of lettuce leaves may vary from dark to light and the quality of the greenness may be modified by a yellowish cast. In addition, anthocyanin may be present, either in an overall distribution or in a pattern of spotting or leaf-margin tingeing. Various combinations of colours and their distributions and of leaf shape, arrangement and folding provide for great variation in appearance of the many cultivars of lettuce.

Lettuce is an annual crop. When the vegetative growth reaches a mature stage, stem elongation occurs and reproductive development begins. Stems vary in length and thickness. A single stem is usually formed but additional stems may form from axillary buds. Cauline, or stem, leaves are usually narrow and clasping at the base.

The inflorescence is a corymbose panicle, composed of many capitula or flower-heads, including a terminal head (Fig. 1.1). Each capitulum consists of several florets, usually from 12 to 20, but as few as seven and as many as 35 (Feráková, 1977). The florets are all ray type, perfect and fertile, and are surrounded by three to four rows of bracts, forming an involucre. Each floret consists of a single, yellow, ligulate petal with five teeth. The lower part is fused as a tube and surrounds the sexual parts.

Each floret has a double carpel, consisting of an elongated style and a divided stigma. There are five stamens; the anthers are fused to form a tube. The flowers open only once, in the morning, remaining open for about 1 h on a warm sunny morning and for several hours when it is cool and cloudy. As the flower opens, the style elongates while the anthers dehisce from within and the shed pollen is swept upwards by the style and stigma hairs.

The ovary is below the corolla. When fertilized it forms an embryo surrounded by nucellar and endosperm tissue and a thin pericarp. The whole is called an achene and is a form of fruit rather than a seed (Fig. 1.2). The achene is topped by a hair-like pappus. Achenes mature about 2 weeks after fertilization. They may be black, grey, white, brown or yellow. The classic discussion of flower and achene development in lettuce is in Jones (1927).

Fig. 1.1. Lettuce panicle showing mature and immature involucres.

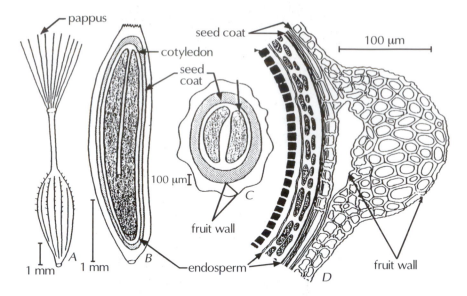

Fig. 1.2. Lettuce achene. A. Entire achene with pappus. B. Longitudinal section. C. Cross-section. D. Portion of wall and endosperm.

Endive is *Cichorium endivia* L. and chicory is *Cichorium intybus* L. These are the two principal species out of nine in the genus (Schoofs and De Langhe, 1988).

Endive is similar to lettuce in many respects. The main differences are in top-growth appearance and in the reproductive structure. The rosette of endive occurs in two forms, both relatively prostrate. One consists of relatively narrow leaves with highly frilled margins. The outer leaves are dark green; the inner leaves are protected from the light and are yellowish or whitish. This form is called endive, narrow-leaved endive or frisée. The other type has broad leaves and is known as escarole or scarole.

Endive is an annual and a seed stalk forms when the rosette accumulates a sufficient number of leaves. The stalk is sparsely leaved and carries sessile flower-heads, borne singly or in groups of two or three (Fig. 1.3). These are contained in involucres and comprise 16–19 florets (Rick, 1953). The flower-head is 3–4 cm in diameter. The flowers are mauve or purplish. The seed-like fruits are achenes and are five-angled and truncated, with several pappus scales.

Wild chicory and the witloof and root-roasting types have relatively long, narrow, strap-shaped leaves that are dark green, forming a broad rosette on a compressed stem. When forced in the dark, leaves are smaller and overlap each other in a tight head or chicon. The leaves contain no chlorophyll and are whitish. Some outer leaves have yellow margins. Another outdoor type is called radicchio, chicorée sauvage or Italian chicory. Radicchio grows in

Fig. 1.3. Inflorescence of endive plant in flowering stage.

several forms: as round compact heads (chioggia), elongated heads (treviso) and cylindrical heads (sugarloaf). They may be dark or light green and may or may not have anthocyanin.

Chicory is biennial and requires vernalization to induce growth of the seed stalk, either naturally by overwintering or by cold treatment (under 5°C) at an early growth stage. The seed stalk, flowers and achenes are similar to those of endive. The flowers are blue, occasionally white or pink, and are about 3.5 cm in diameter. Each flower-head contains 11–28 florets (Fanizza and Damato, 1995).

REPRODUCTION

Lettuce is an obligate self-fertilizing species, primarily because of the structure of the flower. Elongation of the style takes place at the same time as the pollen is released from the inner surface of the fused anther tube. Upon emergence the stigmas and style are covered with pollen grains, which germinate and penetrate the stigmatal surface very quickly. The flower structure and the progress of fertilization mean that the flowers open only once and remain open for a short period during a morning.

Cross-pollination occurs at a very low rate. Thompson *et al.* (1958) estimated the percentage of crossing in an experiment using a dominant gene marker, red colour. The proportion of hybrid plants in the progeny of a green line gave the cross-pollination percentage, which was approximately 1%. The factors that are responsible for the low rate are: (i) the pollen grains are sticky and are not carried by the wind; and (ii) few insects work lettuce flowers. Those that do, thrips and solitary bees, are inefficient because of the flower structure that favours selfing pollen.

Endive is also self-pollinated (Ernst-Schwarzenbach, 1932). An experiment by Rick (1953) showed less than 1% hybridization when paired with chicory.

Despite having a flower structure similar to endive, all forms of chicory are cross-pollinated. The flowers are perfect, but selfing is prevented by self-incompatibility (Stout, 1916). However, the self-incompatibility is not complete; selfing ranges from less than 1% (Rick, 1953) to as high as 20% (Schoofs and De Langhe, 1988). Cichan (1983) found that 70% of collected wild samples of chicory responded to self-fertilization with at least a few seeds. The nature of the self-incompatibility has been uncertain. Pécaut (1962) has suggested sporophytic incompatibility, which is determined by the female genotype. Others have suggested gametophytic incompatibility, determined by specific alleles in both the female and the male gametes. However, Varotto *et al.* (1995) made histological observations on pollinated stigmas and found that the incompatibility reaction was very rapid. This phenomenon, and a genetic study in which the results did not fit a gametophytic system, strongly indicated that a sporophytic system was operating. Two other species in the *Asteraceae*, *Crepis foetida* L. and *Parthenium argentatum* Gray (guayule), also exhibit self-incompatibility due to a sporophytic reaction (Gerstel, 1950; Hughes and Babcock, 1950).

RELATED SPECIES

Lettuce is one of about 100 species of *Lactuca*. Only a few are closely enough related to lettuce (*L. sativa*) to permit crossing. These include *L. serriola* L., *L. saligna* L. and *L. virosa* L. (Fig. 1.4). There are several other forms that may be variants of *L. serriola* or primitive forms of *L. sativa*. These include *L. altaica* Fisch. and Mey., *L. augustana* All., *L. quercina* L., *L. dregeana* D.C. and *L. aculeata* Boiss. and Ky. Zohary (1991) grouped *L. serriola*, *L. aculeata*, *L. altaica*, *L. dregeana*, *L. scarioloides* Boiss., *L. azerbaijanica* Rech. and *L. georgica* Grossh. as the species most closely related taxonomically to lettuce.

Lactuca serriola, common wild lettuce or prickly lettuce, is distributed worldwide, probably having been introduced as a weed on every continent where lettuce is grown. *Lactuca saligna* is found in Europe and western and central Asia, although it has been introduced in North America and Australia

(b)

(a)

Fig. 1.4. Species closely related to cultivated lettuce: (a) *Lactuca serriola*, (b) *Lactuca saligna*, (c) *Lactuca virosa*.

(c)

(Feráková, 1977). *Lactura virosa* is found in western Europe and North Africa. It was probably introduced as a medicinal plant into Europe and possibly also into North America. The compatibility of these species with several others of the genus has been studied, but no successful crosses have been made (Thompson *et al.*, 1941).

Lactuca sativa and its compatible relatives have $n = 9$ chromosomes (Fig. 1.5). This number is most common in the genus. The second most common is $n = 8$. A few North American species have $n = 17$. It is believed that these are amphidiploids of eight- and nine-chromosome species (Thompson *et al.*, 1941).

Compatibility of the close wild relatives with *L. sativa* varies. Crosses between *L. sativa* and *L. serriola* are easily made and are fruitful. For this reason, as well as some morphological similarities, the two forms may be considered to be a pan-species rather than two separate species. In contrast, a cross between lettuce and *L. virosa* produces few viable seeds. The F_1 plants are highly self-sterile and produce very few seeds. However, it is possible to treat the flowers with colchicine and double the chromosome number, thereby

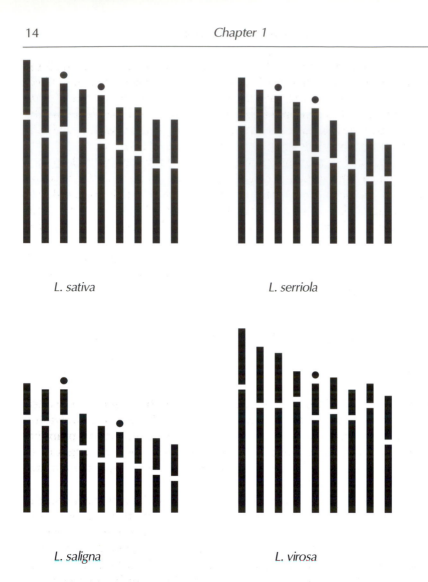

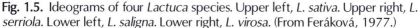

Fig. 1.5. Ideograms of four *Lactuca* species. Upper left, *L. sativa*. Upper right, *L. serriola*. Lower left, *L. saligna*. Lower right, *L. virosa*. (From Feráková, 1977.)

creating an amphidiploid, which will be fertile (Thompson and Ryder, 1961). Crosses with *L. saligna* can be made with less difficulty than with *L. virosa*. Recent work has shown that *in vitro* rescue of embryos can save hybrids between *L. sativa* and *L. virosa* (Maisonneuve *et al.*, 1995). They were also able to use protoplast fusion to regenerate hybrids between *L. sativa* and either *Lactuca perennis* L. or *Lactuca tatarica* (L.) Mey., which are both less closely related to lettuce than is *L. virosa*.

Recent work by Kesseli *et al.* (1991), in which the variation at restriction fragment length polymorphism (RFLP) loci was examined in *L. sativa*, the three related species and two other species, supported the relationships described. Sixty-seven entries were arranged in a dendrogram showing relationship clusters. *Lactuca serriola* appeared to be most closely related to lettuce and the other species were all less close. An accession of *L. augustana* was clustered with a group of lettuce types. This suggests it is closely related to lettuce.

Chicory and endive are quite closely related, as shown in the work by Rick (1953), in which substantial interspecific crossing took place in the direction *C. intybus* × *C. endivia* and less for the reciprocal cross. The difference is due to the self-incompatibility of the former and self-compatibility of the latter. Both species have $n = 9$ chromosomes. There are seven other species of *Cichorium*; all nine have the same chromosome number. Recent work by Vermeulen *et al.* (1994), an analysis of mitochodrial RFLP polymorphisms, suggested that *C. endivia* and *C. intybus* are distinct species, but that *C. spinosum* L. may be an ecotype of *C. intybus*.

EVOLUTION AND CROP HISTORY

Lettuce

Lindqvist (1960b) provided an extensive discussion of the origin of cultivated lettuce. He considered the variation within the four related species and in the group of primitive forms, such as *L. augustana*; he concluded that the primitive forms belonged to *L. sativa*. He advanced three theories on the origin of *L. sativa*: (i) from wild forms of *L. sativa*; (ii) by direct descent from *L. serriola*; and (iii) from hybridization between two species.

The possibility of origin from wild forms of *L. sativa* depends upon the existence of such forms. Lindqvist described a number of primitive forms of *L. sativa*, but claimed that they are not truly wild forms because they were at least semicultivated and have characteristics, such as pointed apex and large seeds, which are different from the closely related *L. serriola*. These forms apparently do not exist in the wild state, but have been shown to be escapes or in cultivation or semicultivation.

The case for direct descent from *L. serriola* rests on the fact that the two species are so closely related. This is based on several groups of traits. The chromosomes are very similar morphologically (Fig. 1.5) and are essentially homologous (Lindqvist, 1960a). Crosses are easily obtained between the two species (Thompson *et al.*, 1941). There is no sterility, regardless of the direction of the cross, and the F_1 is normal and fertile. Comparison of morphological and generative characters and of RFLP variation indicates a close relationship between the two species (Kesseli *et al.*, 1991; DeVries and van Raamsdonk, 1994) (Fig. 1.6). However, DeVries and Raamsdonk consider the differences too great to consolidate them as one species. Kesseli *et al.* found insufficient variation in populations of *L. serriola* sampled to account for the variation in *L. sativa* and suggested that additional populations may have contributed variation.

Lindqvist (1960c) questioned the existence of traits in *L. serriola* favourable enough to induce humans to domesticate it. He concluded that hybridization was required to introduce useful traits and produce the great variety of domesticated types. This hybridization could have been one of three types. One would be the formation of both *L. sativa* and *L. serriola* from a previously existing hybrid population. Another would be hybridization between *L. sativa* and another, unknown, species to form *L. serriola*. This begs the question of the formation of *L. sativa* itself. The third is hybridization between *L. serriola* and another, also unknown, species to form *L. sativa*. No conclusive evidence is presented for any of the hybridization theories.

Origin of *L. sativa* from *L. serriola* by appropriate mutations, leading to forms that lent themselves to domestication, remains the most likely theory. It is reasonable that forms with large seeds and without spines appeared, which then were selected by humans for use and then further modified. The most likely early uses would be for the seeds as a source of oil for domestic use and for the foliage as fodder for animals. Several primitive forms exist, which should be classed as *L. sativa*. Most of these have large seeds that are used

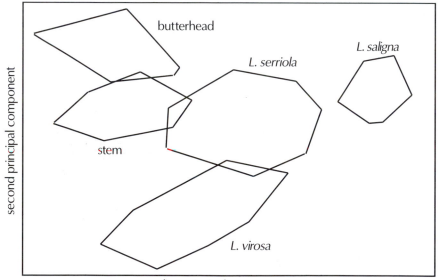

Fig. 1.6. Differentiation among several types of lettuce and species of *Lactuca* by principal-component analysis. Based on 63 generative and vegetative characters of 252 plants. (From De Vries and van Raamsdonk, 1994.)

today for oil. They also have succulent leaves and no spines and could be useful as fodder. They have non-reflexed involucres, which prevents seed loss. They also grow and develop rapidly. These types seem to be the most likely candidates as the first domesticated forms.

The transition to forms edible for humans probably took place in the eastern Mediterranean area, perhaps in Egypt, possibly in the Tigris–Euphrates region. Tomb paintings dating back to the Middle Kingdom of Ancient Egypt, about 2500 BC, show plants considered to be lettuce by several scholars (Keimer, 1924; Harlan, 1986) (Fig. 1.7). Although somewhat stylized, these plants appear to be very similar to stem lettuce or asparagus lettuce. They have long thick stems and narrow pointed leaves. Modern stem lettuces also have the thick stem, but some cultivars have broader, more

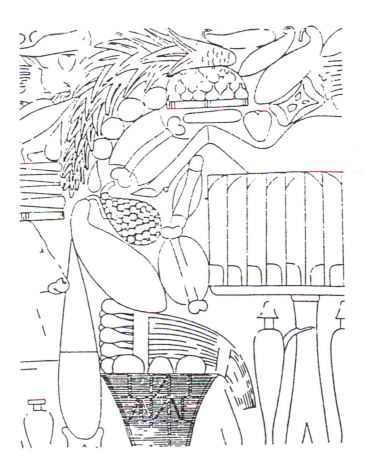

Fig. 1.7. Picture of offerings in rock tomb near Meir, Egypt. Lettuce is at top centre, with pointed leaves. (From Harlan, 1986.)

cos-like leaves. Outside Egypt, lettuce was mentioned as appearing on the tables of Persian kings about 550 BC (Sturtevant, in Hedrick, 1972). In Greece, lettuce was mentioned by Hippocrates in 430 BC and, in Rome, Columella described several types in AD 42. It was mentioned as being in China in the 5th century. In the *Canterbury Tales*, written about 1340, Chaucer described a person as follows: 'well loved he garlic, onions and lettuce'. It was brought to Isabela Island, in the New World, in 1494, on Columbus's second voyage.

Around the Mediterranean basin, cos types predominated. These most closely resemble the stem lettuces and therefore are likely to have developed from the stem types. They occupy the geographical area closest to the likely centre of origin. There is a great deal of variability in the cos lettuces: long and short leaves, flat and erect stature, open and closed heads, variation in texture, red and green types. It seems likely that the leaf or cutting forms and the butterhead, crisphead and Latin types were all selected out of this wealth of variability. The broad, crisp leaf form may have been the predecessor of the Batavia-type crisphead. The modern crisphead or iceberg type was bred from the older Batavia type in recent times (Bohn and Whitaker, 1951).

Until recently, two types of lettuce predominated in Europe. In northern Europe, the butterhead lettuces were most common. These are much more uniform in type than the cos lettuces, varying primarily in degree of greenness. Winter greenhouse types are smaller and poorly headed compared with summer types. In the late 1970s, the types produced and used began to change. Most significantly, crisphead lettuce of the iceberg type began to increase. In the UK and the Scandinavian countries, this type has become the most important one grown and consumed, while usage is increasing substantially in Spain and Germany and less so in France, Italy and the Low Countries. In the Mediterranean area, cos lettuces are most common. The cultivars and landraces of this type share the elongated spatulate-shaped leaf, but vary in leaf dimensions, size of plant, uprightness, degree of greenness, presence of anthocyanin, texture and other traits.

The USA has gone through two major shifts in lettuce usage since the beginning of the 20th century. At that time, a majority of the 21 most popular cultivars were butterhead lettuces (Tracy, 1904) (Table 1.5). In the early part of the century, crisphead cultivars grown in the western USA began to predominate. Lettuce could be grown on large farms, under irrigation, and could be shipped long distances if it had the ability to maintain quality over a period of 10–12 days during shipment and storage. The crisphead type met this last criterion and increased in production at the expense of the other types. By 1955, over 95% of American lettuce was of the crisphead type. About 1980, another shift occurred. The ability of the shipping industry to maintain low temperatures and to transport across the country in as few as 4 days allowed the shipment and handling of leaf, butter and cos types without loss of quality. At the same time, consumers demanded greater variability in

Table 1.5. Major lettuce cultivars grown in the USA in 1904 (From Tracy, 1904).

Butterhead	Leaf	Cos	Crisphead
Big Boston	Black Seeded Simpson	Paris White	Hanson
California Cream Butter	Early Curled Simpson		New York
Deacon	Grand Rapids		Iceberg
Mammoth Black Seeded	Prizehead		
Tennis Ball Black Seeded	White Star		
Golden Queen			
Denver Market			
Hubbard's Market			
Philadelphia			
Reichner			
Silver Ball			
Tennis Ball White Seeded			

lettuce, as in other vegetables and fruit. The proportion of the non-crisphead types has therefore risen to 25–30% of the total production. Similar changes, toward greater variability of the types grown and consumed, are occurring in other parts of the world as well.

Description of horticultural types

Lettuce cultivars exist in a variety of shapes, sizes and colours. Classification into types relies on differences in leaf shape and size, degree of rosette and head formation and less so on colour, stem type and other traits.

Until recently, the most common classification included six types (Rodenburg, 1960). These were: crisphead, butterhead, cos, leaf, stem and Latin. More modern classification includes a primitive group that has been designated oil-seed lettuce (Boukema *et al.*, 1990).

CRISPHEAD   A subtype with large firm heads is designated iceberg in the trade (Fig. 1.8a). This designation should not be confused with the cultivar Iceberg, which actually belongs in the Batavia group described below. The first true iceberg type was the cultivar Great Lakes, developed in the USA in 1948 (Bohn and Whitaker, 1951). Typical American cultivars are large, weighing about 1 kg, with six or seven outer leaves. The heads are firm to hard, giving only slightly to pressure when mature. The plant passes through a rosette stage first. The early leaves are elongated; each successive leaf increases in broadness, until they become broader than long when mature. At about 10–12 leaves, a leaf becomes cup-shaped, enclosing the later leaves, which leads to the development of a spherical head. The head continues to enlarge and fill from the inside, until market maturity is reached. If the head is not harvested, it then goes into the reproductive stage, with the elongation of the seed stalk. Leaf texture varies from highly crisp (Great Lakes group) to less crisp (Salinas–

Vanguard group). The outer leaves are either bright green or dull green and the interior colour varies from almost white to creamy yellow.

A second crisp subtype includes the Batavia lettuces. These crisp lettuces originated in Europe and are most commonly found there (Fig. 1.8b). In England, Webb's Wonderful is a Batavia type, as are the early crisp types in the USA: Iceberg, Hanson and two transitional types closer to the iceberg subtype, New York and Imperial. There are a large number of cultivars in France, often with the word Batavia as part of the name. They form and develop in a similar manner to the iceberg types, but are more variable in final shape and are smaller and less dense. They weigh about 500 g when mature.

BUTTERHEAD Butterhead lettuces also originated in Europe and became the most popular ones grown there (Fig. 1.8c). There are two general subtypes in Europe, based upon season of growth. Summer, outdoor types are larger, weighing about 350 g. They are well filled at maturity and slower bolting than the winter types, which are smaller, less well filled and may weigh 150–200 g. In the USA, there are also two subtypes, based on appearance and size. The

(a)

(b)

Fig. 1.8. (a) Head of crisphead lettuce, iceberg type cv. Salinas. (b) Head of crispbread lettuce, Batavia type cv. Pierre Benitez. (c) Head of butterhead lettuce, cv. Midas.

(c)

Boston subtype is larger, lighter in colour and has a softer texture. The Bibb subtype is smaller and darker green. Butterhead leaves of all types are relatively thin, with an oily soft texture. The colour of outer leaves is lighter than most iceberg-type lettuces. The interior colour is yellowish.

COS   This type is also known as romaine or roman. Cos lettuces have been traditionally grown in the Mediterranean basin. The name comes from an island (Kos) in the eastern Mediterranean near Turkey. There is great variation of appearance. The colour ranges from yellowish to dark green. The leaves are elongated (Fig. 1.9). The rosette tends to have an upright stature, although some are quite flattened. Upright cultivars form a loaf-shaped head after the rosette stage and may be either open or closed at the top. Leaf texture is rather coarse. A high proportion of leaves are green, because of the relatively open nature of the head. Interior leaves are yellowish. Cos heads may weigh as much as 750 g.

LEAF   Leaf or cutting lettuce cultivars are often grown in home gardens and in covered structures (Fig. 1.10a). Leaf lettuces also vary greatly in appearance. Leaves may be broad, as in crisphead types, or lobed, resembling oak leaves, or elongated, like cos lettuce. They may form an upright or flattened rosette. Leaves may be entire at the margin or highly frilled. They may be green or yellow in varying shades and may have anthocyanin in

Fig. 1.9. Field of cos lettuce under drip irrigation.

varying amounts. Leaf cultivars form a rosette of leaves, which may be closely bunched or relatively flat and open. The bunched types may form rudimentary heads under low temperature conditions.

LATIN   These are also called grassé lettuces (Fig. 1.10b). They are of European origin, but are also grown in South America and on small areas in the USA. Their upright stature and elongated leaves resemble cos lettuce, but the leaves are shorter. Leaf texture resembles Bibb-type butterhead lettuce; it is soft but somewhat thickened. Some cultivars have rather leathery leaves.

STEM   These are also called stalk or asparagus lettuces (Fig. 1.11). As the name implies, they are grown for the stem, which elongates as the rosette is forming and which is thick (5–7 cm in diameter). The leaves are long and may be quite narrow or as broad as cos leaves. The leaves are removed and the stems are eaten raw or cooked. Stem lettuces are found in Egypt and Middle Eastern countries; judging from the Egyptian tomb paintings, they are quite ancient. Stem lettuces are also common in China. They have been known there since the 6th century AD and may have been carried overland from the Middle East at an early time.

OIL-SEED   This is a group of lettuces that progress through the rosette stage very rapidly and bolt early (Fig. 1.12). They have been classified as *L. serriola* or with other appellations, but have traits which indicate they are primitive forms of *L. sativa*. Most especially, they often have seeds that are about 50% larger than those of other lettuces. These seeds are pressed to produce an oil for domestic use, such as cooking. This is done today, but the practice is apparently very ancient and may have been the first use for *Lactuca* by humans. The plants produce long, narrow leaves and bolt and flower rapidly. It

(a)   (b)

Fig. 1.10. (a) Head of green leaf lettuce, cv. Salad Bowl. (b) Heads of Latin lettuce, cv. Sucrine.

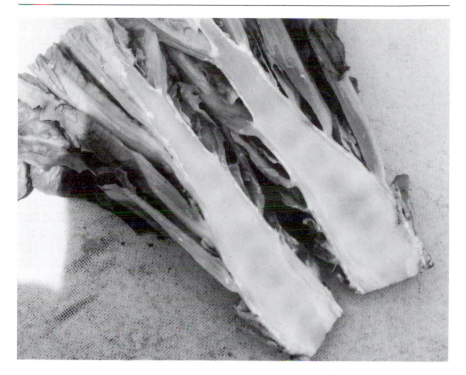

Fig. 1.11. Longitudinally cut thick stems of stem lettuce, Balady type.

is the only group identified here that has not been thoroughly domesticated by the development of landraces and cultivars.

Endive

The origin of endive is somewhat obscure. According to Sturtevant (Hedrick, 1972), it first appeared in India. However, he noted that other unnamed authors fixed its origin in Sicily. It was used as a salad vegetable at 'a very early period' in Egypt and was also known by the Greeks and Romans, which lends some support to a possible Sicilian origin. In Rome it was mentioned by Ovid, Pliny and Columella. Endive appeared in England as early as 1548. It was first mentioned in a seed catalogue in the USA in 1806, but it must have come to the Americas at an earlier date. The broad-leaved type, known as escarole, seems to be older than the narrow-leaved type. First mention of the narrow-leaved type was in the 13th century.

Description of horticultural types
Endive comes in two forms, broad-leaved and narrow-leaved.

Fig. 1.12. Plant of PI 251246, an oil-seed lettuce, in flowering stage.

BROAD-LEAVED   This type is called escarole or scarole. It forms a semiopen head 30–33 cm in diameter, with relatively broad leaves and slightly frilled margins (Fig. 1.13a). The outside leaves are green; inside leaves are creamy white to yellow. The outer leaves are quite bitter; the inner leaves have less chlorophyll and are not as bitter.

NARROW-LEAVED   This type is called endive or frisée. The leaves are narrower than the escarole type and are much more frilled (Fig. 1.13b). There is some variation in the degree of frilliness, from moderate to extreme (très fine). The heads are looser and larger in diameter than the escarole type. The proportion of bleached leaves is smaller and the degree of bitterness is greater than with the escarole type.

Chicory

Chicory is widely distributed over western, central and southern Europe, North Africa and portions of Asia. It probably originated in the Mediterranean basin (Schoofs and De Langhe, 1988). The ancient Egyptians, Greeks and Romans used the green leaves as a salad vegetable and the roots for medicinal purposes. It was harvested in the wild before the various specialized forms were developed. The first mention of cultivation was in 1616, in Germany. Near the end of the century, it became a cultivated plant in England.

In 1775, two French physicians discovered that chicory roots could be dried, roasted and ground for use as a substitute or a flavouring for coffee. In many countries, this use continues to be popular.

The origin of the witloof form has been attributed to a M. Bréziers, of the State Botanical Garden in Brussels, who in about 1850, discovered some 'forgotten roots' that had sprouted in the dark, forming white leaves. The leaves were elongated and loose and the product was known as barbe de capucin. When larger roots were used and forced under a cover of sand or soil,

(a)

(b)

Fig. 1.13. (a) Escarole (broad-leaved endive) in late rosette stage, cv. Altis. (b) Endive (frisée type) in late rosette stage, cv. Green Curled Ruffec.

the tightly headed chicons were formed. Witloof was developed from the Magdebourg cultivar. Belgium became the cradle of culture. Witloof became very popular in France and the Netherlands and those countries developed their own industries, in 1873 and after 1913, respectively. Witloof chicory is also known as Belgian endive, Brussels endive and French endive.

Italy is the home of the greatest variety of non-forced chicory types, including red and green types, rounded or elongated heading types and rosette types; appearance varies greatly within each type.

Description of horticultural types

ROSETTE TYPE   This type, which is cultivated as a green vegetable, is similar in appearance to the wild form. There are two forms, one with elongated and divided leaves and one with entire leaves. The midrib may be green or red. The rosette is open and leaves continue to form until the reproductive phase. This type has other uses. The root may be dried, roasted and ground to be used as a coffee substitute or processed industrially to produce fructose sugar. Rosette-type chicory is forced to produce witloof chicons. This type is used for animal fodder in New Zealand and is being tested for that purpose in the USA.

WITLOOF TYPE   This is the forced stage of the rosette type. The apical bud (chicon), which is produced in sand or soil or in the dark, is composed of white or yellow-tipped leaves folded over each other, often very tightly, and it is usually pointed at the upper end (Fig. 1.14). There are also red cultivars. Chicons may range in length from 9 to 20 cm to meet extra or class one standards and may be 2.5–8 cm in diameter.

HEADING TYPES   These are red, red and green, red and white or green chicories grown for field production of edible leaves. The red and variegated types (radicchio) may be spherical or elongated (Fig. 1.15). The heads usually weigh about 450 g. The spherical types are about 12 cm in diameter. Green types have an elongated, fairly compact head (sugarloaf), about 35 cm long, or form a short-leaved, loose head.

Fig. 1.14. Chicons of witloof chicory forced in hydroponic growing room. (Courtesy H. Bannerot.)

Fig. 1.15. Mature plant of radicchio type of chicory.

GENETICS AND BREEDING

Breeding is the means by which new genes for yield, disease resistance and other useful traits are incorporated into crops. It enables crops to cope better with their environment, such as insects, diseases and stress problems. It also leads to improvement in colour, taste, head size and shape and other traits that make them desirable food products. In turn, breeding is made more efficient by knowledge of the inheritance of these traits and the manner in which they interact with the environment and with each other.

GENETIC STUDIES IN LETTUCE

Lettuce is not a model species for plant genetic studies comparable to tomato, maize or *Arabidopsis*. Nevertheless, significant progress has been made in the identification of both genes and molecular markers, and in their physical and functional relationship to each other, thus contributing to the development of a genetic map. These include genes useful to breeding.

Over 90 genes have been identified in lettuce. These may be grouped in categories: disease and insect resistances, leaf morphology and colour, floral morphology and colour, seed and fertility characteristics and growth and development.

Disease Resistance

Of great interest and importance are resistances to diseases and insects. Downy mildew is an economically important disease of lettuce in most areas in which it is grown. After the first inheritance study by Jagger and Whitaker (1940), information on the genetic basis for resistance accumulated in a piecemeal manner before I.R. Crute and A.G. Johnson proposed a gene-for-gene hypothesis specifying the genetic relationship between the disease organism and the lettuce host (Crute and Johnson, 1976; Johnson *et al.* 1977, 1978). Single dominant alleles in the host plant, lettuce, confer resistance to isolates of the disease organism, *Bremia lactucae* Regel, expressing the corresponding avirulence allele with the same number as the lettuce allele. The lettuce alleles are designated *Dm-1*, *Dm-2*, etc. Virulence is recessive (Norwood *et al.*, 1983b). For example, *Dm-1* confers resistance to all forms of the fungus except those containing the virulence allele *avr-1*. The same relationship holds for *Dm-2* and *avr-2*, etc.

Much of the recent work on the genetics of resistance to downy mildew has been driven by the need to find new sources of resistance. This need is due to two factors. One is the development of resistance in some isolates of the fungus to metalaxyl, a major chemical used in controlling the disease. The other is due to the ability of *B. lactuceae* to overcome resistance genes. These phenomena will be discussed in the breeding section in this chapter.

The early work by Crute and others identified 11 probable resistance genes, which were referred to as R factors, in the absence of definitive genetic information. As each factor was identified as a specific gene pair through inheritance studies, they were symbolized as *Dm* alleles. These and subsequent studies by other workers identified the following *Dm* alleles: *Dm-1*, *Dm-2*, *Dm-3*, *Dm-4*, *Dm-5/8*, *Dm-6*, *Dm-7*, *Dm-10*, *Dm-11*, *Dm-13*, *Dm-14*, *Dm-15* and *Dm-16* (Farrara *et al.*, 1987). Additional R factors have been identified, up to R-30 (Bonnier *et al.*, 1994) (Table 2.1). In addition, new resistances have been identified in the related *Lactuca* wild species *L. saligna*, *L. virosa* and *L. serriola* (Netzer *et al.*, 1976; Norwood *et al.*, 1981; Farrara and Michelmore, 1987).

The *Dm* genes are race-specific. Some work has been done to search for non-race-specific resistance. Crute and Norwood (1981) identified 81 cultivars and lines with no known *Dm* alleles or R factors and subjected them

Table 2.1. *DM* genes or R factors of lettuce cultivars and lines used as differentials for identifying downy-mildew-virulence phenotypes of *Bremia lactucae*. (From Farrara *et al.*, 1987; Michelmore, personal communication.)

Cultivar or line	Gene or factor	Cultivar or line	Gene or factor
Cobham Green	0	Hilde	12
Lednicky	1	Empire	13
UCDM2	2	UCDM14	14
Dandie	3	PIVT1309	15
R4 T57E	4	LSE18	16
Valmaine	5/8	Kinemontepas	10, 13, 16
Sabine	6	Alpha	1, 5, 8
LSE57/5/Mesa 659	7	Target	11, 5/8, 13
UCDM10	10	Mariska	R18
Capitan	11		

to field infection. Twenty-one of these showed some level of field resistance. Further tests identified three cultivars, Iceberg, Batavia Blonde de Paris and Grand Rapids, as showing a significantly high level of resistance as measured by a reduced percentage of plants with symptoms, fewer leaves with symptoms and a lower mildew score. Crosses of Iceberg and Grand Rapids with susceptible cultivars showed that inheritance was probably quantitative. However, it was reasonably easy to select progeny equivalent to the resistant parents, so the number of genes controlling disease reaction is probably relatively low (Norwood *et al.*, 1983a, 1985). Seedling resistance of these cultivars was not obviously correlated with the level of field resistance (Norwood and Crute, 1985).

Genetic and molecular studies have shown that several *Dm* genes occur in clusters (Hulbert and Michelmore, 1985; Farrara *et al.*, 1987; Kesseli *et al.*, 1994). The largest group includes *Dm-1*, *Dm-2*, *Dm-3*, *Dm-6*, *Dm-14*, *Dm-15* and *Dm-16*. A second group includes *Dm-5/8* and *Dm-10* and a third contains *Dm-4*, *Dm-7* and *Dm-11*, while *Dm-13* maps to a fourth linkage group that may be loosely linked to the Dm-7/Dm-11 group.

Lettuce mosaic is a virus disease that has been a serious problem in most or all lettuce-growing areas worldwide. Resistance was identified by Von der Pahlen and Crnko (1965) in the cultivar Gallega, and by Ryder (1968) in three plant introductions from Egypt. The resistance in Gallega is recessive and was named *g* (Bannerot *et al.*, 1969); the resistance in the Egyptian lines is also recessive and was named *mo* (Ryder, 1970). It was at first believed that the genes were identical, but differential reactions to isolates of lettuce mosaic suggest that they may be different alleles of the same locus. Systemic infection occurs in *g/mo* plants, but the disease develops slowly and is manifested by small chlorotic areas rather than complete mottling of the leaf surface. An additional gene for mosaic reaction, dominant for resistance to a rare isolate of the virus, has been identified in the cultivar Ithaca (Pink *et al.*, 1992a).

Corky root is a bacterial disease of lettuce. Resistance was first identified by Dickson (1963) in several plant introductions. Resistance was later confirmed by Sequeira (1970), who developed the first resistant cultivars. A recessive allele, *cor*, was identified as conferring resistance (Brown and Michelmore, 1988).

Three other disease resistances have been assigned specific gene identifications. Bidens mottle is a virus disease controlled by a single recessive allele *bi* (Zitter and Guzman 1977). Resistance to powdery mildew, a fungus disease, is controlled by a single dominant allele, *Pm* (Whitaker and Pryor, 1941; Robinson *et al.*, 1983). Turnip mosaic is a virus disease, with resistance controlled by a single dominant allele, *Tu* (Zink and Duffus, 1970). This gene has been mapped near *Dm-5/8* and several other markers (Robbins *et al.*, 1994).

Insect Resistance

Resistance to root aphid has been identified in several cultivars. Resistance in the cultivars Avoncrisp, Avondefiance and Lakeland is controlled by a single dominant allele (Ellis *et al.*, 1994). This is contrary to previous work indicating that resistance was due to maternal effects (Dunn, 1974). The gene is linked to the largest *Dm* cluster.

The nature of resistance to foliar aphids is less clear. Eenink and Dieleman (1983) showed that resistance to the leaf aphid *Nasonovia ribis-nigri* is due to either a dominant or partially dominant allele from *L. virosa*. Resistance to another leaf aphid, *Myzus persicae*, is partial and probably quantitative (Eenink and Dieleman, 1977).

Disease and insect resistances are discussed further in Chapters 7 and 8, respectively.

Leaf Colour and Morphology

A number of genes for leaf morphology and colour have been identified. A gene with three identified alleles governs leaf lobing: u^+ produces a pinnatifid leaf, as in *L. serriola*, u^o produces the oak-leaf shape and u produces non-lobed leaves (Lindqvist, 1958; Robinson *et al.*, 1983). Several genes affect the degree of serration on the leaf margins. All are recessive.

Lettuce leaf colour varies with the amount of chlorophyll, governing the degree of greenness, and with the amount of anthocyanin, governing the presence, pattern and distribution of red colour.

Many different shades of green exist in lettuce, as well as a number of relatively deleterious mutants causing various patterns of chlorophyll deficiency. Several of these types have been identified genetically. All those showing a reduction in the amount of green colour (chlorophyll) present are due to recessive alleles. Table 2.2 shows known genes for changes in green colour and in anthocyanin differences.

The classic anthocyanin paper is that by Thompson (1938). He proposed a complementary gene pair controlling presence or absence of anthocyanin (*CcGg*) and a multiple allelic system controlling the pattern and distribution of colour. If the dominant allele of both complementary genes is present, red colour will appear. Any other combination gives green. The *R* allele gives a general distribution of red over the surface of the leaf; R^s gives a red-spotted leaf; and R^t gives a reddish tinge on the leaf margin. Lindqvist (1960c) identified another allele, R^{rb}, for red-brown colour, as well as two additional genes, one for colour intensification and the other for reduction of colour with the age of the plant. The anthocyanin genes also affect the distribution of colour in the stems, involucres and flower petals.

Table 2.2. Genes for various forms of leaf-colour differences in lettuce and related species. (From Robinson *et al.*, 1983; Ryder, 1983; Ryder, 1996.)

Gene name	Symbol	Description
Changes in greenness		
albino-1, 2, 3	al-1, 2, 3	Lethal, no chlorophyll
alboxantha	ax	Yellow, except leaf base
chlorophyll-deficient-1 to 7	cd-1 to 7	Yellow, mottled variations
calico	cl	Variegated leaves
golden	go	Chlorotic leaves, golden flower
golden yellow	gy	Golden-yellow leaves
light green	lg	Light green leaves
virescent	vi	Yellow leaves turning green
dappled	dap	Fine mosaic, white, yellow, green
sickly	si	Chlorotic, necrotic, slow growth
apple green	ag	Light silvery green
shiny green	sg	Glossy green
Anthocyanin variation		
Basic genes	C, G	Both dominant, permit red colour
Red	R	Red overall
Spotted	R^s	Red spots
Tinge	R^t	Marginal red tinge
Red-brown spotted	R^{bs}	Brownish-red young leaves
intensifier	i	Intensifies effect of *R* allele

Floral Genetics and Male Sterility

Lettuce flowers are normally yellow. Three recessive mutants are known that change the colour from yellow to pale yellow, golden and salmon, respectively. Seed colour also varies. One gene controls black versus white; the latter is recessive. A recessive allele at another locus produces brown colour, while a recessive allele at a third locus produces yellow.

Involucre development can be affected in two ways. A recessive allele produces a swollen or plump immature involucre. Wild lettuce, *L. serriola*, produces an open or reflexed involucre at maturity, which is dominant to the closed or erect type produced by most cultivated lettuces.

Lindqvist (1960c) identified three complementary loci for male sterility: *ms-1*, *ms-2* and *ms-3*. Ryder (1963, 1967, 1971) identified three types of male sterility. The first is controlled by a dominant–recessive type of epistasis, such that *ms-4Ms-5* produces sterility. The second is a recessive, *ms-6*, which has two pleiotropic effects, small flowers and slightly twisted leaves. The third is a dominant, *Ms-7*, which also produces folded petals. A recent paper by Curtis *et al.* (1996) described genomic male sterility in *Agrobacterium* transformed plants of the cultivar Lake Nyah. The transgenic plant expressed a glucanase enzyme in the anther locule, which caused malformed pollen grains.

Growth and Development

Three aspects of growth and development have been investigated for their genetic bases: head development, flowering time and dwarfism. Heading is a complex process, affected by both genetics and environment. The overall process is probably controlled quantitatively (Durst, 1930; Lewis, 1931). Single genes have also been identified. Bremer and Grana (1935) crossed heading with non-heading types and named a gene *Kopfbildung*, with the recessive *k* producing the heading type. Later, this was confirmed by Pearson (1956), who studied a non-heading rogue in the cultivar Imperial 456. Lindqvist (1960c) found two additional genes, the dominant forms of which conferred non-heading.

Flowering time and bolting are related characters. Bremer and Grana (1935) identified a single gene controlling earliness of bolting. They named the recessive allele *tagneutral (t)* for the genotype that was independent of day length. The dominant *T* caused bolting to occur only under long days. Ryder (1983, 1988) identified two partially dominant alleles, *Ef-1* and *Ef-2*, for early flowering in crisphead-type lettuce. The double dominant genotype reduces flowering time by two-thirds, as much as 100 days, over the double recessive. These two genes are quantitatively dependent upon day length. The earliest types go directly into the bolting stage, after forming a few leaves but no rosette (Fig. 2.1).

Waycott *et al.* (1995) identified four loci affecting growth. Each recessive allele generates a dwarf phenotype, producing plants ranging in height from 3 to 30 cm. These anomalies were created by mutagenic treatment of early-flowering lines. Flowering time is not affected by the dwarfism.

Linkage and Mapping

The genetic map for a species can be derived using two approaches. One is to make many diverse crosses to obtain linkage data between pairs of genes. The other is to identify linkages by detailed analysis of many or few crosses using molecular methods. The overall map is generated from both types of information.

Relatively little traditional genetic linkage information is available for lettuce. Several linkage pairs have been determined but only a few linkage groups with more than two loci have been established. Lindqvist (1960c) found that four genes, *light green* (*lg*), *golden yellow* (*gy*), *hearting* (*h*) and *intensifier* (*i*), are linked in that order. Ryder (1983) described a five-gene group: *endive* (*en*), *white* (*w*), *virescent* (*vi*), *fringe* (*fr*) and *male sterile-6* (*ms-6*), in that order.

In developing a genetic linkage map from deoxyribonucleic acid (DNA) markers, linkage groups for disease-resistance loci were established (Kesseli *et*

Fig. 2.1. Expression of early-flowering genes in plants of same age. From left, Ef-1Ef-1EF-2Ef-2, Ef-1Ef-1ef-2ef-2, ef-1ef-1EF-2Ef-2, ef-1ef-1ef-2ef-2.

al., 1994). One comprised *Dm-10*, *Dm-5/8* and *Tu* (*turnip mosaic resistance*) in that order. A second included eight loci: *Ra* (*root aphid resistance*), *Dm-15*, *Dm-6*, *Dm-16*, *Dm-14*, *Dm-2*, *Dm-3* and *Dm-1*, in that order. A third contains *Dm-4*, *Dm-7* and *Dm-11*, with the order undetermined.

Molecular Genetics and Tissue Culture

A detailed genetic map has been constructed by Kesseli *et al.* (1994), consisting of markers identified by restriction fragment length polymorphism (RFLP), random amplified polymorphic DNA (RAPD) and isozyme analyses. All the loci have so far coalesced into 13 major linkage groups and four minor groups. A number of loci remain unlinked. Higher-density maps are being constructed from other crosses, and nine linkage groups, corresponding to the number of chromosomes, will be established by R. Michelmore and associates.

Cloning of disease-resistance genes in lettuce is based upon a map-based approach. This method depends upon establishment of the physical and genetic location of a gene close to molecular markers on either side. It led to the location of genetic markers in the region of *Dm-1* and *Dm-3* (Fig. 2.2) and in the region of *Dm-11* (Paran *et al.*, 1991; Paran and Michelmore, 1993). The *Dm-3* locus was further characterized by Anderson *et al.* (1996).

Transformation is the insertion of a gene into a host plant, often using the plant-pathogenic bacterium *Agrobacterium tumefaciens*, minus its pathogenic property. Transformation of lettuce was first done successfully on the cultivar Cobham Green (Michelmore *et al.*, 1987). Although transformation can now

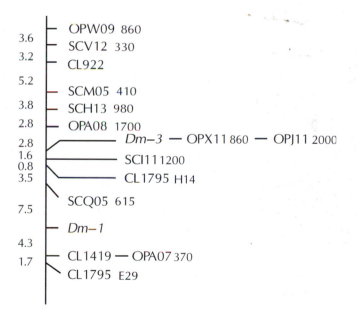

Fig. 2.2. A portion of the lettuce linkage map, showing the *Dm-1* and *Dm-3* region of linkage group 2. Constructed from the F$_2$ of the cross Calmar × Kordaat. Markers detected by a complementary DNA (cDNA) clone (CL), an RAPD primer (OP) or a SCAR primer pair (SC). Genetic distances on left in centiMorgans. (From Paran and Michelmore, 1993.)

be accomplished routinely, transgenic lettuce may express the transferred gene weakly or not at all, as compared with other species.

Tissue culture has become a routine procedure in many species, including lettuce. Doerschug and Miller (1967) were first to regenerate shoots from several tissues. Engler and Grogan (1984) described regeneration of lettuce plants from protoplasts. Michelmore and Eash (1986) summarized the literature and the various protocols used by different workers.

Cytogenetics

Gates and Rees (1921) first showed that lettuce (*Lactuca sativa*) has a chromosome number of $n = 9$. T.W. Whitaker and colleagues, in a series of papers starting with Whitaker and Jagger (1939), explored the chromosomal and genetic relationships among various species and found that *L. sativa* can only be crossed with *L. serriola*, *L. saligna* and *L. virosa*. *Lactuca sativa* and *L. serriola* are closely related, but it was difficult to obtain viable hybrids in crosses between *L. sativa* and *L. virosa* until R.C. Thompson made this cross successfully in 1938, by doubling the chromosome number of the F_1 and obtaining a fertile amphidiploid (Thompson and Ryder, 1961). It was found in later work that viable hybrids could also be generated in diploid material.

Quantitative Genetics

Although many traits must be inherited quantitatively in lettuce, little published work is available on biometrical studies that define quantitative parameters. Durst (1930) found that leaf length and width, plant height, flowering time and plant habit were probably quantitatively inherited. In a study on the inheritance of fasciation, Eenink and Garretsen (1980) found that the regression of F_3 means on F_2 individuals showed narrow-sense heritability ($h^2 = 0.4$). In a diallel study of nitrate content, Reinink (1991) showed that additive gene effects accounted for most of the variation among the parents.

GENETIC STUDIES IN ENDIVE AND CHICORY

Relatively few genetic studies in endive and chicory have been recorded. Tesi (1968) found apparent heterosis for leaf number and leaf length in endive crosses. The inheritance of these characters was complex.

Rick (1953) crossed Radichetta chicory with White Curled endive in a study to determine whether the two types should be considered the same or separate species. The endive was self-pollinated, while the chicory produced nearly all hybrids due to its self-incompatibility. Nine characters showed dominance of the chicory type or intermediate status in the F_1. In the F_2, flower colour was inherited as a single gene trait, with the blue of chicory dominant to the mauve of endive. Internode length was also a single gene trait, with long dominant to short. Peduncle length and fasciation were also apparently controlled by single genes. Other characters, including hairiness, bract width, width–length ratio of rosette leaves, degree of inflorescence, tertiary branching and terminal internode length, were probably quantitatively inherited.

Garibaldi and Tesi (1971) crossed chicory and endive to study the inheritance of *Alternaria* resistance. They found that the resistance of chicory was dominant and a single gene was involved.

Olivieri (1972) did a quantitative inheritance study in a cross between a radicchio and a witloof chicory. He found additive effects for leaf number, plant width and seed-stalk emergence.

Chicory has $n = 9$ chromosomes. An ideogram of the species was developed by DuJardin *et al.* (1979) (Fig. 2.3).

The use of tissue culture in regenerating plants of *Cichorium intybus* has been summarized by Schoofs and De Langhe (1988). The procedures include bud, shoot and root regeneration, protoplast culture, haploid induction and somatic embryogenesis. Haploid plants can be created by pollination with *Lactuca tatarica* L. or *Cicerbita alpina* Walbr. (Doré *et al.*, 1996).

BREEDING AND CULTIVAR DEVELOPMENT

Lettuce

Breeding strategy is dependent upon the breeding system inherent in the crop species. Lettuce is a natural inbreeder due to the structure of the flower and absence of self-incompatibility genes. This limits the practical types of breeding strategies to three: pedigree breeding, single-plant or mass selection

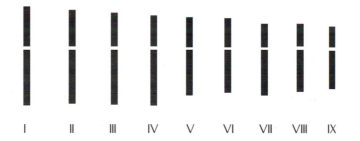

Fig. 2.3. Ideogram of chromosomes of *Cichorium intybus* (From Dujardin *et al.,* 1979).

and back-crossing. Hybrid F_1 production is more appropriate to naturally cross-pollinated species. Although hybrids have been developed with some self-pollinated crops, the technique has yet to be exploited in lettuce.

Crossing technique

Each lettuce floret has a monadelphous stamen tube surrounding the style (Fig. 2.4). On the morning that the flower opens, the styles elongate as the anthers dehisce at their internal surfaces. Thus the style and stigmatic surfaces are covered with pollen as they emerge from the stamen tubes. Normally, the pollen will quickly germinate, penetrate the stigma and effect self-fertilization. To prevent this, several different methods of manipulation are possible.

One can emasculate the florets by lifting off the stamen tube with a fine pair of forceps before the style elongates (Ernst-Schwarzenbach, 1932). Pollen from another parent can then be applied to the washed stigmas. A higher percentage of crossing can be ensured by clipping off the anthers and removing any remaining pollen with a stream of air (Pearson, 1962).

Most lettuce breeders use a method developed by Oliver (1910), or a modification of his method. If pollen is washed off the emerging stigmas as they begin to part, a relatively high percentage of crossing can be obtained. Earlier or later washing will decrease the percentage of crosses and increase the proportion of selfing. If the flower is continually washed with a fine mist from the beginning of the emergence process, a very high percentage of

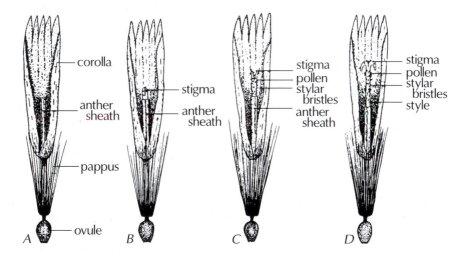

Fig. 2.4. Stages in opening of lettuce flower. A. Stigmas have not emerged from anther sheath. B. Stigmas at early emergence stage. C. Stigmas beginning to part; ideal stage for crossing. D. Stigmas curled backwards; pollination is complete, past crossing stage. (From Thompson, 1938.)

crossing, approaching 100%, can be obtained (Ryder and Johnson, 1974). It is easier and just as effective to wash the flowers with a directed stream, two or three times during the emergence process. In either case, the flowers should be dried with a stream of air before applying the foreign pollen. Finally, a combination of clipping and washing will result in 100% crossing (Nagata, 1992).

Breeding methods

The pedigree method, selection and back-crossing are the three principal breeding strategies used in lettuce breeding. The method used depends upon the goal of the programme.

Pedigree breeding is the preferred method when there is a need to combine favourable characters from two or more parents and to eliminate unfavourable characters from each parent. Pedigree breeding is often used when the parents are cultivars and each contains a combination of favourable and unfavourable traits. The F_2 population is very important. It should be as large as can possibly be handled. This generation has the greatest amount of variation resulting from the cross, which increases the probability of finding plants with maximum useful and minimum non-useful traits. With each generation of selection and inbreeding, uniformity within each family increases. After several generations, the breeder may then elect to grow the selected lines in field trials under commercial conditions, to compare them with existing cultivars. If one or more of the lines is considered sufficiently superior to the existing cultivars, after several tests under varying environments, the breeder may then release seed of the line(s) as a new cultivar or cultivars for use by commercial growers.

When a new cultivar is released while still containing some variability or when it has been in existence a long time and increased several times, leading to an accumulation of mutational changes, selection within the cultivar may be practised. Selection produces variants of the original cultivar, which may now be considered new, even though they are similar to the original. This is a practice much used by seed companies to develop new cultivars from publicly developed landmark cultivars, which are usually released to all seed companies with lettuce programmes. This gives each company the opportunity to develop and sell cultivars different from those sold by other companies.

Certain cultivars may be outstanding in most respects but lack a specific character controlled by one gene, such as resistance to a disease. If the allele conferring the resistance is in another cultivar, landrace or primitive or wild type, that allele can be transferred to the outstanding cultivar by a process called back-crossing.

Back-crossing begins with a cross between an F_1 plant and the outstanding parent. The latter becomes the recurrent parent with repeated back-crossing between it and the progeny of the first and subsequent crosses.

With each back-cross the proportion of genes from the non-recurrent, donor, parent is reduced by half. After six back-crosses, over 99% of the genome is from the recurrent parent. The desired allele and a small section of chromosome on either side of the allele remain from the donor parent.

The back-cross process may be improved by reduction of the number of generations required to reach a given state of resemblance to the recurrent parent. This can be achieved by selection for molecular markers near the desired gene to reduce linkage drag. Similarly, but probably less cost-effectively, markers unlinked to the desired gene can also be selected. In both cases, the breeder has the opportunity each generation to select those plants which have a minimum proportion of genomic material from the donor parent (Michelmore, 1995).

Another modification of the back-cross method involves the use of early-flowering genes to speed up the process by reducing generation time (Ryder, 1985). Genes *Ef-1ef-1* and *Ef-2ef-2* affect flowering time such that the double dominant reduces flowering time by about two-thirds (from 140–150 days for most crisphead lettuce to 45–50 days during long days) (Ryder, 1983, 1988). Under the same conditions, the double heterozygote flowers in 65 days. If the early-flowering alleles are combined with the desired gene for transfer, each back-cross generation will require about half the time from planting to seed production compared with normal-flowering lines. Therefore, the time for the entire procedure can also be reduced by about one-half (Table 2.3). *Ef* genes can be removed by selfing and selecting for the *ef* alleles.

There has been some interest in the development of F_1 hybrids for lettuce, primarily as a security measure. However, there are inherent difficulties in creating F_1 hybrids, such as near absence of pollen transfer by wind or insects and a very low number of seeds per hand pollination. Therefore, to date, there are no lettuce hybrids.

Table 2.3. Back-cross sequence for transfer of lettuce mosaic resistance allele from an early flowering line (EF) to the cultivar Prizehead (PZ). Days to flower columns show expected elapsed time for four back-crosses (BC_1–BC_4) with standard back-cross procedure (PZ) and for accelerated procedure (EF). (From Ryder, 1985.)

Cross	Plant date		First flower		Days to flower	
	PZ	EF	PZ	EF	PZ	EF
PZ x EF	19 Nov. 82	21 Jan. 83	15 Apr. 83	20 Mar. 83	147	58
BC_1	6 May 83	10 Jun. 83	8 Aug. 83	15 Aug. 83	97	66
BC_2	29 Jul. 83	9 Sep. 83	10 Nov. 83	20 Nov. 83	104	72
BC_3	21 Oct. 83	20 Jan. 84	17 Mar. 84	25 Mar. 84	152	70
BC_4	6 Apr. 84	1 Jun. 84	23 Jul. 84	2 Aug. 84	110	64

Breeding goals

There are four general goals of lettuce breeding. These are: (i) horticultural improvement; (ii) resistance to diseases, insects and stress problems; (iii) uniformity of maturity; and (iv) adaptation to specific environments.

Under horticultural improvement, objectives include optimum size, weight and yield; improved colour; good head or rosette conformation; and improved taste and texture.

Size and weight needs vary depending upon the type of lettuce and the nature of the market. Crisphead lettuces in the USA are expected to be large and heavy, while in Great Britain and other European countries smaller heads are desired. Butterhead lettuces are universally smaller than crisphead types but vary in size among subtypes.

Yield of lettuce can be measured in two ways: by the number of units or by weight. Within the industry in the western USA, yield is measured in cartons, which average 24 heads each and weigh about 25 kg. In this sense, the yield of a field is closely related to the number of plants in the field, since one plant equals one unit of yield. Plant spacing is uniform throughout the planting area; therefore, the maximum possible yield is the same for all fields. Most statistics recorded by various agencies show yield as weight per unit area. Since each plant is a yield unit the former method is more informative. Yield is a function of earliness and uniformity.

There are three aspects of colour variation that are important: intensity of greenness, distribution and intensity of anthocyanin and interior colour, especially of the heading types. Dark green exterior colour, whether dull or bright, is usually preferred in crisphead lettuce. Most butterhead cultivars are lighter green, sometimes ranging to yellow green. Cos lettuces are usually dark green, but in southern Europe and the eastern Mediterranean basin, they tend towards lighter green or yellow green. Leaf lettuces have a considerable range of colour, ranging from light to dark green. Anthocyanin is a light-regulated phenomenon and relatively few crisphead, butterhead and cos cultivars are red, since the interior leaves show no red colouring. A large proportion of leaf lettuces, with their open rosettes, are red, which may range from red-tinged leaf margins to full red over most or all of the leaf surface. Colour shade may be red, red-purple, red-brown or pink.

The most desirable shape for crisphead lettuce is spherical. Heads that are pointed at either end are considered undesirable since they are more difficult to pack in a carton. Butterhead lettuces are also usually preferred rounded, although the top of the head is slightly flattened in outdoor types. Greenhouse types have less well-formed heads. Cos heads are elongated and may be either relatively open at the top or slightly closed, with the apices of the leaves bent inward. Leaf lettuces have rosettes of various shapes. The rosette may be relatively upright and some cultivars may tend to form rudimentary heads, especially in cool weather. This latter tendency is considered undesirable. The rosette may be open and the appearance will vary depending upon leaf type: broad or narrow, frilly or smooth, lobed or unlobed.

Lettuce has a rather mild taste regardless of type. The main goal of the breeder is to avoid bitterness and select towards sweetness. Crisphead and cos types should have a crisp, crunchy texture, while butterhead lettuces are soft with an oily texture. Leaf lettuces range from crisp to soft. Latin lettuces are intermediate between butterhead and cos type.

A major set of goals for lettuce improvement is the breeding of cultivars resistant to diseases, insects and stress problems. The specific resistances desired depend upon the area where the cultivar will be grown; however, many of the problems are ubiquitous.

Among the virus diseases, lettuce mosaic and big vein are found in many areas, and resistances to them are included as goals in many breeding programmes. Lettuce infectious yellows and lettuce chlorosis are problems in desert production areas. Cucumber mosaic and broad bean wilt are found in the eastern USA, tomato spotted wilt primarily in Hawaii and lettuce necrotic yellows in Australia.

The two most common fungal diseases are downy mildew and sclerotinia drop. Many programmes for resistance to downy mildew are in existence, both in the USA and in Europe. Although sclerotinia is a ubiquitous problem, few resistance programmes exist. Stemphylium is a problem in Israel and resistance is a goal there. Resistance to anthracnose is a goal in one programme in California.

Corky root is a bacterial disease found in many growing districts in the USA, as well as in Europe and Australia. Resistance breeding is practised in several programmes.

Although there are many insect species that cause problems in lettuce, there are breeding programmes for resistance to only a few. These are the lettuce root aphid in the USA and Europe, three species of foliar aphids in the Netherlands and the sweet potato whitefly and pea leafminer in the western USA. In future, as the use of chemicals continues to decline, breeders may consider programmes for resistance to several species of lepidopterous worms, as well. Insect-resistance breeding and genetics are discussed in Chapter 8.

Among the stress problems, tipburn and premature bolting can occur anywhere lettuce is grown, whenever high temperature is likely to occur. Although the genetics of these problems is not known, and no organism is involved, most breeders try to select against the disorders. Pink rib and rib blight cause unsightly damage to lettuce. Cultivar differences in susceptibility are known and may eventually be incorporated into breeding programmes.

In addition to the field problems mentioned, there are postharvest diseases and disorders of lettuce. These include a fungus disease, botrytis, and bacterial slime from several organisms, as well as environmentally induced disorders, such as russet spotting. Each of these is potentially amenable to improvement through breeding.

Uniformity refers to the ability of a cultivar, in a given environment, to reach maturity so that all or most of the plants can be harvested in one cutting. The fewer times required to cut all the plants in a field, the less costly and more efficient the process becomes. Lack of uniformity is more of a problem with crisphead lettuce than with the other faster-maturing types. Together with earliness of maturity, uniformity maximizes potential and actual yield.

Lettuce is produced in a variety of environments at all times during the year. Temperature, day length, soil type, moisture and other environmental factors all vary from location to location and season to season. The ability of lettuce cultivars to grow well in different environments varies widely. A few crisphead lettuce cultivars, such as Salinas and Great Lakes 659, are widely adapted and can perform relatively well in most lettuce-growing areas. Others have much narrower environmental niches. In general, the breeder's goal is to develop cultivars adapted to fairly specific environments. Subsequent testing in other areas will disclose whether they are more widely adapted. Non-crisphead lettuces, which are faster growing and have less stringent heading requirements, tend to be relatively widely adapted.

Protected lettuce

The breeding goals for lettuce in tunnels, greenhouses and growing rooms include the fulfilment of some specific needs. Cultivars must be adapted to low, artificial, energy inputs (light and temperature) and sometimes to increased input of carbon dioxide. Nitrate accumulation in leafy vegetables is greater in winter environments, in which protected lettuce is grown. Therefore, reduction of nitrate accumulation must be a breeding goal. Downy mildew, botrytis and other fungus problems may require greater attention than for outdoor lettuce. The nutrient film technique (NFT) is an ideal means for transferring root-infecting organisms, such as the big vein agent, and resistances to these problems should be an important goal.

History of lettuce breeding

The origin of most lettuce cultivars in use in the latter part of the 19th century and the early part of the 20th century is relatively obscure. Breeding and selection were undoubtedly practised by individual growers and seed companies for many years, as indicated by the many cultivars in existence at the turn of the century. The earliest recorded formal public breeding programme began in California in 1923. At that time, I.C. Jagger was transferred by the US Department of Agriculture (USDA) to the San Diego area in southern California to develop cultivars resistant to a disease of unknown cause called brown blight. This disease caused severe losses in lettuce fields in California, which, in the early part of the 20th century had become the principal source of lettuce for shipment to all parts of the USA. Jagger selected resistant survivors in fields of the New York cultivar group, then the principal crisphead cultivars grown, and from these developed Imperial 2, Imperial 3 and Imperial

6. Later he made crosses to incorporate resistance to downy mildew along with the brown blight resistance. A large number of Imperial cultivars were developed and soon replaced the group of New York cultivars (Jagger *et al.*, 1941).

The Imperial group dominated US production through the 1930s and 1940s and continued to be produced in significant volume through the early 1950s. In 1941, the cultivar Great Lakes was released by the USDA and the Michigan Agricultural Experiment Station (Barrons and Whitaker, 1943). This cultivar came out of the Imperial group from a cross with Brittle Ice, which contributed increased solidity, so that Great Lakes was the first true modern crisphead. At least 60 cultivars were developed, and this group was the principal type grown through the mid-1970s in the USA. Great Lakes cultivars were also grown in other countries. In the latter part of the Great Lakes period, the cultivar Calmar and a subgroup developed from it occupied a major place. Calmar was bred for downy mildew resistance by the University of California and released in 1960 (Welch *et al.*, 1965).

Ross C. Thompson started a lettuce-breeding programme for the USDA in 1938 at Beltsville, Maryland. The most significant breeding accomplishment from this programme was the development of four cultivars, Empire, Merit, Climax and Vanguard, which have dominated the desert lettuce-production areas from the early 1960s to the present. Vanguard was a special accomplishment, since it was the first cultivar developed from a cross between lettuce and the wild species, *L. virosa* (Thompson and Ryder, 1961).

A cross between a Vanguard-like breeding line and Calmar led to development of the cultivar Salinas by the USDA in 1975 (Ryder, 1979a). The Vanguard–Salinas type has dominated lettuce crisphead production world-wide since then. (In Europe, Salinas is known as Saladin.)

Lettuce breeding for the production areas in the eastern USA has been done mostly at Cornell University in New York and at the University of Florida. Beginning with the cultivars Oswego and Fulton, developed at Cornell, a series of cultivars have come from programmes in those states for production on organic soils.

Outside the USA, the principal public breeding programmes have been in England and the Netherlands. In England, early research at the National Vegetable Research Station (Horticultural Research International) culmin-ated in the release of two downy-mildew-resistant cultivars, Avoncrisp, a crisphead type, and Avondefiance, a butterhead. Subsequent work there has also focused on downy mildew resistance, leading to the development of the gene-for-gene hypothesis, specifying the relationship between the lettuce and the fungus, *B. lactucae.* At the Glasshouse Crops Research Institute, the main emphasis was on the development of butterhead cultivars tolerant to low light and temperature for winter production under cover.

Similar research on downy mildew resistance and winter production has been carried out in the Institute for Horticultural Plant Breeding in

Wageningen, the Netherlands (now part of the Centre for Plant Breeding and Reproduction Research). In addition, two other major programmes have developed. One is the breeding of lettuce cultivars requiring lower nitrogen input and with lower content of nitrate nitrogen. The other programme emphasizes research on the inheritance and breeding of resistance to foliar aphids.

Work on root aphids has been carried out in both countries. In recent years, as the production of crisphead lettuce has increased in several countries, breeding of cultivars of this type has also increased, but primarily in private breeding programmes.

Public-sector lettuce breeding has traditionally concentrated on the creation of landmark cultivars addressing specific difficult problems, often by bringing in genes from wild species and non-adapted cultivars and landraces. Seed companies involved with lettuce-seed production have had lettuce-breeding programmes whose primary objective has been the modification of existing cultivars for specific production needs. Recently, however, financial support for public programmes has decreased and private companies have expanded their research to include disease-resistance programmes and other more basic aspects of breeding.

Over the years, there have been several notable breeding achievements. These include the development of the several types and subtypes of lettuce, nearly all accomplished by unknown breeders in the centuries before plant breeding existed as a formal science and art.

The iceberg subtype of crisphead lettuce is the only type developed and documented in the modern era. The cultivar Great Lakes was developed by T.W. Whitaker and released in 1941, jointly with the Michigan Agricultural Experiment Station (Barrons and Whitaker, 1943). It was a departure from the then currently grown Imperial cultivars in having a very firm large head, very crisp leaves and wide adaptability; it was the first cultivar of the type and nearly all modern crisphead cultivars derive from it.

Another important accomplishment in the history of lettuce breeding is the development of the gene-for-gene hypothesis explaining the relationship between the lettuce host and the downy-mildew-inducing organism *B. lactucae* (Crute and Johnson, 1976). This hypothesis proposed that each major mildew gene in lettuce had a corresponding virulence gene in the organism. A specific *Dm* allele in the lettuce cultivar conferred resistance to all virulence alleles in the organism except one with the same number. Subsequent research has confirmed this hypothesis (see beginning of chapter and Chapter 7).

Future of breeding

The array of present-day cultivars of the various types are products of standard breeding techniques and have been developed primarily in the public sector. There are indications that lettuce breeding in the future may change in several aspects. New techniques will be added to those being used now. Changes will take place in the public and private sectors.

Molecular biology should have a significant impact. Molecular markers will add to breeders' ability to pinpoint the location of useful genes and should make identification of the most useful genotypes easier. This will be true even for quantitatively inherited traits. Transformation using *Agrobacterium* spp. or other means will enable breeders to transfer genes not easily accessible with standard breeding techniques. Cell and tissue culture may permit rapid uniform multiplication of desirable materials, such as F_1 hybrids. Alternatively, manipulation of the cultured materials will allow the creation of additional variability, some of which may be useful in breeding.

The future of public-sector plant breeding is uncertain. In recent years, there has been a shift in emphasis away from the development of cultivars to molecular biology, genetics and germplasm manipulation and enhancement. At the same time, private companies have consolidated and added staffs in both breeding and molecular biology. The consequences of these changes for lettuce breeding cannot be accurately predicted. However, the possibility exists that there will be a shift in emphasis from actual breeding programmes leading to breakthrough cultivars, which have been the hallmark of public-sector breeding, to greater emphasis on molecular studies, germplasm evaluation and early stages of breeding. The burden of cultivar development may fall on companies doing breeding, which may focus on relatively short-term programmes, leading to quicker release of marketable cultivars. A possible consequence of such changes may be a reduced overall emphasis on innovative breeding.

Endive

The breeding system of endive is similar to that of lettuce. Ernst-Schwarzenbach (1932) found that most flower-heads were self-fertile and averaged over 11 seeds per flower. Crosses between endive types were also fertile but averaged fewer seeds. Rick (1953) found that endive plants under screenhouse protection were almost as fertile as plants in the field. Nearly all seeds on endive plants were from self-pollination, while most seeds on chicory plants were from crosses.

Endive is a minor crop compared with lettuce and chicory. Therefore, there is very little work reported on breeding improvement. Nearly all the breeding work has been done by seed companies, and the results have been reported in their seed catalogues. In the USA, nearly all the cultivars are variations of Green Curled (narrow leaf) and Full Heart Batavian (escarole). In France, endives have been selected for greater variation in fineness of leaf, degree of blanching and seasonal adaptation.

Very little disease-resistance breeding has been reported for endive.

Provvidenti *et al.* (1979) reported on resistance to turnip mosaic in chicory and proposed transferring the resistance to endive and escarole. Zitter and Guzman (1977) found little or no resistance to bidens mottle virus in tests of several endive cultivars and Plant Introduction (PI) accessions. They did report finding resistance in chicory cultivars.

Chicory

Incompatibility

Although chicory flowers look very much like endive flowers, the breeding systems are different. Chicory is primarily cross-pollinated, as it is largely self-incompatible (Stout, 1916). However, self-incompatibility is not complete; self-fertile plants are found sporadically among hybrids between self-incompatible parents (Stout, 1917). It is possible to select and inbreed self-compatible lines, with minimum inbreeding depression (Eenink, 1980).

In a series of papers, Eenink (1981a, b, 1982, 1984) discussed compatibility and incompatibility in witloof chicory and their importance in plant breeding studies. Pollen germination was most active between 17 and 20°C. Seeds did not necessarily form even when pollen germinated on the flowers. It was suggested that incompatibility is due to a single-locus sporophytic system with varying dominance relationships. Selection for high general and specific combining ability can be practised. Double and triple pollinations, including use of day-old pollen, can induce selfing in incompatible forms.

Witloof breeding

Traditional cultivars of witloof chicory were developed by mass selection. In Belgium, most cultivars were selected by farmers for their own use and sale to neighbours. These were given the name of the selector, such as Penninckx, Christaens or Mueninck, with the addition of an adjective indicating early, mid- and late maturity. Dutch and French seed companies gave true cultivar names to their releases.

The practice of breeding F_1 hybrid cultivars stems from the pioneering work by J.A. Huyskes, of the Institute for Horticultural Plant Breeding, Wageningen, the Netherlands. This work grew out of a perceived need to force chicon development in the absence of soil cover. In the traditional, highly labour-intensive method, soil cover was necessary to compress and maintain the compactness of chicon growth. Huyskes (1963) conducted a series of experiments comparing forcing conditions, treatment of the roots, relation of core length and earliness and selection for improved head compactness and quality during forcing without soil cover. He found that existing cultivars did not force well without cover, but that cooling of the roots improved the quality. He found that core length was an excellent indicator for earliness, and that it could be used as a criterion for selecting for earliness and uniformity. He developed a core meter for this purpose (Fig. 2.5). He also found that, in a breeding programme, he could select plants that formed compact heads without cover, a major factor in modernizing the witloof industry.

H. Bannerot (Station de Génétique et d'Amélioration des Plantes, Versailles, France) released the near-F_1 hybrids Flambor, Bergere and Zoom. Zoom, in particular, became highly popular because of its uniformity. It was also found by Lips (1976) that Zoom was the best performer among several cultivars forced without cover. Many hybrid cultivars have been developed, mostly by private breeders, and are in production.

There are several goals in the breeding of witloof cultivars. The original farmer-developed cultivars were populations selected for production of uniform tight heads of acceptable commercial size and shape. Now the goals are similar, but the traits are incorporated in F_1 or near-F_1 hybrids for forcing without cover. Some types are bred exclusively for tray production, and others for either tray or soil-planted production.

Cultivars are tailored for a specific season of production. Early cultivars have longer core lengths than late cultivars and this is used as a criterion of selection for early-, mid- and late-season cultivars (Huyskes, 1963). Those designated for early-season forcing require a shorter period of vernalization, store less well and grow rapidly during forcing. Cultivars for late-season forcing are harvested later, store better, require more chilling and grow relatively slowly during forcing. Late types produce more compact chicons.

Red colour in the forcing types became a goal recently. Bannerot and de

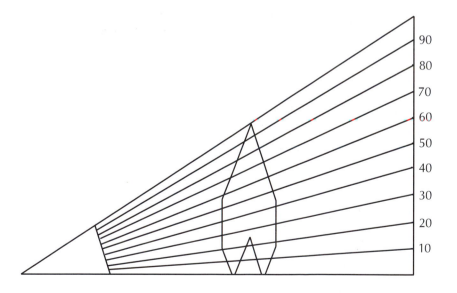

Fig. 2.5. Core meter for measuring relative core length in chicon to estimate earliness (From Huyskes, 1963).

Coninck (1976) crossed the red Italian cultivar Rosso di Verona with several witloof types. Red witloof cultivars are now on the market.

Other breeding objectives include tight closure of the tops of the chicons, tolerance to internal browning, resistance to premature bolting, reduced bitterness and good presentation quality with regard to colour, shape and uniform size.

Industrial-use breeding

Breeding for industrial use is a relatively new enterprise. As demand for certain agricultural crops has declined, the need for rotation with new crops has been investigated. The use of chicory roots for roasting to make a coffee substitute or ameliorant is an old practice and limited in potential. However, the roots contain inulin, a polymeric chain of fructoses ended by a glucose molecule, which is a precursor of fructose and oligofructoses (Fig. 2.6). There is increasing interest in using chicory for sugar production. As chicory does not yet have the competitive production capacity of sugar beets or starch-containing plants, such as potatoes and maize, the major breeding goal is increased potential for sugar production. Fructose yield in grams is highly correlated with root weight and moderately correlated with

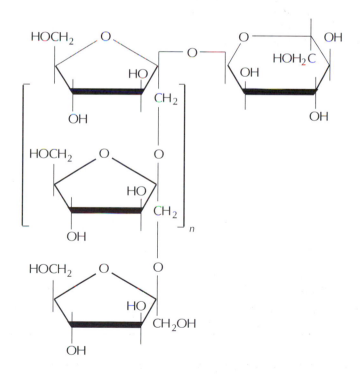

Fig. 2.6. Diagram of inulin molecule.

percentage fructose (Coppens d'Eeckenbrugge *et al.*, 1989).

Other goals include the development of cultivars with resistance to premature bolting, greater adaptation for mechanized harvesting, tolerance to diseases, root storage quality, improved germination and improved root yield. In a comparison of cultivars used in 1972 and in 1992, Leroux (1994) showed that the later cultivars were improved in bolting resistance, total dry matter, percentage of dry matter and total root yield.

Desprez *et al.* (1994) proposed a maternal selection scheme, combined with *in vitro* cloning, to produce improved populations, which can then be tested in hybrid combinations (Fig. 2.7). A population is grown and selections are made based on slow bolting, disease tolerance, root form and other favourable phenotypic traits. Samples are taken from final selections for laboratory testing of dry-matter content. Selections with high dry matter are then grown for seed and also preserved *in vitro*. The selections are tested for germination, reselected for phenotypic traits and the corresponding *in vitro* cultures of the best selections are regenerated to produce seed, which is massed and placed in another population for a second round of selection. After several rounds of selection, the best populations are intercrossed and evaluated for best hybrid combinations.

Frese and Dambroth (1987) suggested, after testing 49 accessions of both root and leaf chicory, that the latter form might have insufficient variability in root size and sugar content to contribute to rapid breeding progress in root chicory. Hybrids between the two types might produce greater variability.

Louant *et al.* (1978) proposed that achene characteristics could be combined in a discriminant function to identify and purify cultivars. Varietal identification and purity of F_1 hybrids can also be ascertained with the use of RAPD techniques (Bellamy *et al.*, 1996).

History

Wild chicory is found in most parts of the world, including Europe, North Africa, North America and the temperate regions of Asia. It is probably the prototype for similar cultivated types, such as Radichetta. A great variety of chicory types have been developed from the early forms; these include green- and red-leaf types, which form heads of various shapes, from spherical to elongated; this group is also known as chicorée sauvage and radicchio. Other specialized derived types are witloof chicory (Belgian endive) and the root types used for a coffee substitute or for production of inulin-based sugars.

As is typical of most crops that are cross-pollinated, early breeding of chicory was conducted by mass selection. Mass selection allows the preservation of superior types in relatively heterozygous genotypes that maintain vigour. However, development of F_1 and near-F_1 hybrids has become the norm for the leaf, witloof and industrial-type chicories. Inbred lines are crossed to form hybrids that are productive and uniform.

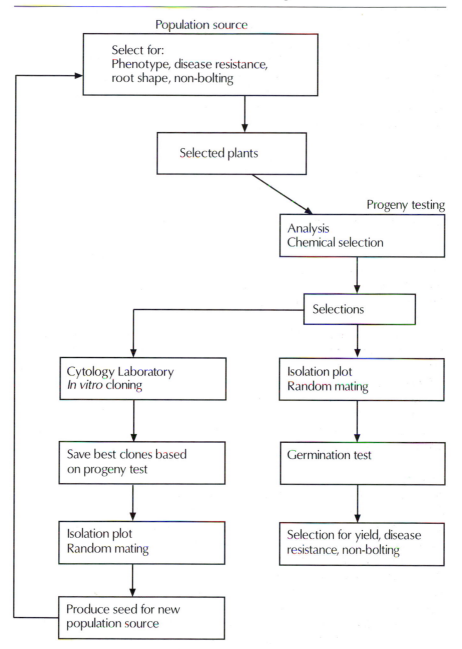

Fig. 2.7. Flow-chart representation of selection scheme, combining progeny testing and *in vitro* cloning for development of improved lines of industrial chicory. (From Desprez *et al.*, 1994.)

GERMPLASM RESOURCES

Germplasm is the life blood of plant breeding. It supplies the new genes needed to enable crops to be competitive against various pests and to maintain and increase yield, quality, derived products and important cosmetic traits. In the broadest sense, germplasm consists not only of introduced wild species and landraces, but also the genetic and breeding stocks developed in active programmes, as well as commercial cultivars. The products of biotechnology, including DNA clones, tissue-culture calli and cell cultures, can also be considered germplasm.

There are five activities that must be undertaken to properly exploit the usefulness of germplasm. These are exploration and collection, storage and maintenance, evaluation, distribution and breeding research. The last activity makes use of germplasm accessions in breeding programmes, in genetic studies and in other applications.

The beginning of the process is exploration for and collection of accessions of wild species and landraces, as well as the collection of new cultivars from various seed companies and the saving of breeding and genetic stocks from existing programmes. Lettuce has been collected over the years from most European countries, the Mediterranean basin, the Middle East and western Asia, India, China and Japan. Endive and chicory collection has been on a more limited scale.

Collected seed must be kept in storage under conditions supportive of long-term viability. For many species, including lettuce, endive and chicory, a cool, relatively dry environment is required. The minimum conditions are 5°C at 40–50% relative humidity. Storage in a freezer at −18°C will keep seed viable up to 25 years, or longer, depending upon the initial condition of the seed. Storage in liquid nitrogen will provide very long-term viability but is expensive, cumbersome and not useful for working collections. The International Board for Plant Genetic Resources recommends hermetic storage at −18°C or less and 5% seed moisture content.

The collection should be evaluated for useful traits. These may include morphological characters, yield, other performance traits and resistances to diseases, insects and stress. Those materials with the favourable traits can then be used in a breeding programme or other study.

The ultimate goal of long-term storage is preservation for possible future use. The most important use is the incorporation of the favourable traits identified by evaluation into breeding lines and eventually into new cultivars. However, the information derived by evaluation can also be useful in genetic, pathological or physiological studies. For example, a researcher may wish to compare yield potential of resistant and susceptible lines with and without the stress of the disease.

There are five major collections of *Lactuca* in the USA and four in Europe (Table 2.4). The largest of these is at the US Agricultural Research Station, Salinas, California.

Table 2.4. Lettuce and *Lactuca* species germplasm collection centres.

Name	Location	Materials
Horticultural Sciences Department, Cornell University	Geneva, New York	Cultivars, landraces, breeding lines, genetic stocks, wild species
Lactuca Genetic Resources Collection, University of California (UC) Davis	Davis, California	Cultivars, landraces, breeding lines, genetic and molecular stocks
Lettuce Collection, Centre for Genetic Resources	Wageningen, the Netherlands	Cultivars, landraces, wild species
N.I. Vavilov Institute of Plant Industry	St Petersburg, Russia	Cultivars, landraces
Research and Plant Breeding Institute	Olomouc, Czech Republic	Cultivars, landraces
USDA–ARS Regional Plant Introduction Station	Pullman, Washington	Landraces, wild species
USDA *Lactuca* Germplasm Collection	Salinas, California	Cultivars, landraces, breeding lines, genetic stocks, wild species
US National Seed Storage Laboratory	Fort Collins, Colorado	Cultivars
Vegetable Gene Bank, Horticulture Research International	Wellesbourne, UK	Cultivars, landraces, wild species

There are three moderate-sized collections of the *Cichorium* species in the USA. One is at the USDA Regional Plant Introduction Station at Ames, Iowa. The others are at the US Agricultural Research Station, Salinas, California, and the Horticultural Sciences Department, Cornell University, Geneva, New York. Major collections in Europe are located at the Centre for Genetic Resources in Wageningen, the Netherlands, and the Horticultural Research International, Wellesbourne, UK.

PHYSIOLOGY OF GERMINATION, GROWTH AND DEVELOPMENT

Physiological studies in lettuce, endive and chicory include research on seed dormancy and factors influencing germination, rosette and head (heart) formation and the seed-stalk elongation and flowering processes. Physiological disorders may occur at various growth stages, particularly near maturity and in postharvest environments. These aspects will be discussed in this chapter. Effects of modified atmospheres are discussed in Chapter 5. The nature of seed quality, seed coatings, after-ripening and seed priming are discussed in Chapter 6.

GERMINATION PROCESS IN LETTUCE

The lettuce seed is actually an achene, defined as a dry, indehiscent, single-seeded fruit. The lettuce seed consists of an embryo with two cotyledons surrounded successively by layers of endosperm, integument and pericarp. It germinates, following water imbibition, by means of radicle extension. A description of the structure of the lettuce achene and the influence of various internal conditions and environments on germination has been published by Borthwick and Robbins (1928). A number of authors have since described various aspects of germination and certain chemical changes occurring during the process.

Ikuma and Thimann (1964) described the germination process in three physiological phases. They investigated the effects of light and heat together, and also the effect of oxygen (O_2), by conducting the experiments in either a nitrogen (N_2) or O_2 atmosphere. They analysed the changes taking place in each of the three phases at a standard temperature (25°C) and variations from the standard.

1. Preinduction phase. Water is imbibed in this phase, which takes about 1.5 h at the standard temperature. Water uptake rate increases with increasing temperature. Lack of oxygen has no effect. Sensitivity to red light increases, at an increasing rate, as the temperature is raised. At very high temperatures, subsequent germination in the dark is inhibited, but the inhibition may be modified by exposure to red light.

2. Induction phase. Sensitivity to red or far-red light is at a maximum at this time. Exposure to red light for 1 min is sufficient to initiate germination. Temperature changes and presence or absence of oxygen have no effect during this phase.

3. Postinduction phase. This phase takes about 9 h at the standard temperature. A reaction requiring oxygen takes place immediately after exposure to red light. This phase is also temperature responsive, with inhibition occurring at 35°C. This is also the escape phase, during which exposure to far-red light cannot reverse the red light effect.

Pollock and Manalo (1971) compared lettuce seeds germinated at 20°C and at 25°C under three levels of moisture stress in lighted growth chambers. At the lower temperature, there was little effect of moisture stress on germination. At the higher temperature, however, germination decreased with increasing moisture stress. They suggested that moisture stress as well as light are important for germination response.

Wurr and Fellows (1987) showed that germination decreases with decreasing water potential. They found that seeds would germinate normally if the soil moisture content was at 0 bar for 24–48 h. Bradford (1990) modified an equation describing cell growth to describe initial growth of the radicle during imbibition. Thus, germination time-course curves could be generated in terms of the water potential. He showed that the rate of germination of lettuce seeds (cv. Empire) increased linearly with the turgor of the embryo.

The relative roles of the endosperm and the seed-coats in delaying or preventing germination is uncertain. Endosperm cell-wall autohydrolysis at the micropylar end is required for radicle extension to take place. Endo-β-mannanase activity may play a role in this process (Dutta *et al.*, 1997). Wurr *et al.* (1987a) showed that greater force was required to overcome the constraint of the pericarp than of the endosperm.

Complete digestion of the endosperm and transfer of its nutrients to the embryo occur in the germinating seed; lipids are the primary food sources (Park and Chen, 1974). Lipids are utilized for respiration and synthesis of amino acids, and are partially converted to sucrose.

Burdett (1972) showed that ethylene production is reduced when seeds are imbibed at 30°C and that this is the mechanism for inhibition of germination. Van Staden (1973) proposed that conversion of cytokinins from water solubility to butanol solubility is a key event in germination, while Daines *et al.* (1983) implicated the induction of phenylalanine ammonia-lyase (PAL) as another key event.

The two main environmental factors that directly affect germination are light and temperature.

Influence of Light

Flint and McAlister (1937) showed that red or white light promoted germination, while far-red light or the absence of light inhibited germination. The specific action spectra were identified by Borthwick *et al.* (1952, 1954) as 660 nanometres (nm) for promotion and 735 nm for inhibition (10 Å = 1 nm). They proposed a light–plant-pigment reaction, and showed that the reaction was endlessly reversible and that the last exposure is the controlling one. The plant pigment is called phytochrome and it can take two forms, Pfr when exposed to red light and Pr when exposed to far-red light. When the ratio of the Pfr form to total phytochrome is high, after exposure to red or white light, germination is promoted. When the ratio is low, after far-red exposure or darkness, germination is inhibited.

The effects of light can be modified in various ways. Certain chemicals can overcome the inhibitory effect of far-red light or darkness. Gibberellin (Kahn *et al.*, 1957) promotes germination under far-red light, while gibberellin plus ethylene has a synergistic promotional effect (Burdett and Vidaver, 1971). The polyamines putrescine and spermidine stimulated germination under dark conditions (Sinska, 1988). On the other hand, ancymidol, an inhibitor of gibberellin synthesis, inhibited germination under red light (Gardner, 1983).

Under relatively low irradiance levels with white light, the effect of red light is dominant over the effect of far-red light. However, at high levels of irradiance, the effect of far-red light becomes dominant and germination is inhibited. This is independent of the effect of temperature (Gorski and Gorska, 1979).

Glutamine synthetase activity is related to germination (Takeba, 1983). The data suggested that inactivation of glutamine synthetase may lead to thermodormancy in seeds of cv. New York.

Ellis *et al.* (1989) found a response to photon dose, which is the product of photon flux density, the amount of light per unit area and photoperiod length, in four *Asteraceae* species. In lettuce, there was no response at lower temperatures, but at 25°C there was a direct linear response until light exposure became continuous, when a negative response occurred (Fig. 3.1).

Inoue and Nagashima (1991) located the photoperceptive site on the lettuce seed by irradiation with microbeams of red light at points along the length of the seed, after removal of the pericarp. If the seed length is divided into eight sections lengthwise, the receptive site is located in the second section from the radicle end, which corresponds to the hypocotyl (Fig. 3.2).

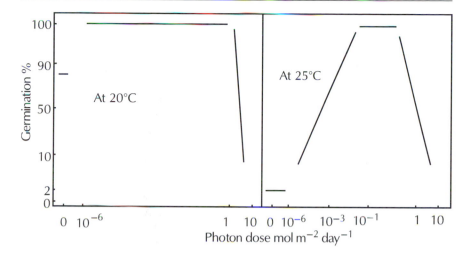

Fig. 3.1. Photon-dose response of lettuce seeds. Germination at constant temperatures of 20°C and 25°C at increasing photon dose by exposure to white light. (From Ellis *et al.*, 1989.)

Influence of Temperature

The second important influence on germination is temperature. Germination of lettuce seed takes place at a range of temperatures, with an optimum of 18–21°C. At or above 26°C, germination may be inhibited. This is called thermo-dormancy. The extent of inhibition varies with cultivar and type. As a group, crisphead cultivars were found to be less subject to thermal dormancy than cos and butterhead cultivars (Gray, 1975). The highest temperature at which 50% of the seeds germinated ranged from 25.7 to 30.5°C for 16 butterhead cultivars and from 28 to 32.8°C for four crisphead cultivars and was 31°C for two cos cultivars (Table 3.1).

Thermodormancy can be overcome by the addition of thiourea (Thompson and Horn, 1944), kinetin (Smith *et al.*, 1968) or ethrel (Harsh *et al.*, 1973). There are also synergistic effects. Gibberellin and kinetin together enhanced germination at high temperature more than either alone (Haber and Tolbert, 1959), while carbon dioxide (CO_2) in the presence of ethylene enhanced germination at 35°C (Negm *et al.*, 1972). Suzuki (1981) found that after-ripening also had an effect on germination. Germination increased with increased after-ripening time up to about 3 years. Subsequent to that, deterioration took place, even at 15°C, and germination decreased to zero.

Coons *et al.* (1990) compared germination of ten cultivars at four temperatures and five sodium chloride (NaCl) concentrations. Crisphead cultivars differed from each other and from Grand Rapids as the temperature increased from 20°C to 35°C and the differences increased with increasing salt levels.

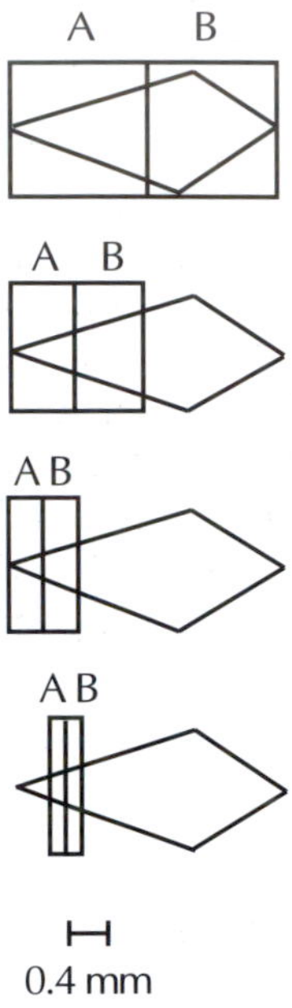

| | Germination percentage irradiated seed portion | | | |
	A	B	Whole seed	Far-red control
	55	15	70	5
	42	5	47	0
	4	72	55	10
	35	16	75	5

Fig. 3.2. Effects of red light on germination induction at different locations along length of decoated seeds. Root tip at left. (From Inoue and Nagashima, 1991.)

Table 3.1. Highest temperatures allowing 50% germination after 7 days. Comparison of butterhead, cos and crisphead lettuce cultivars. (From Gray, 1975.)

Type	Cultivar	Temperature (°C)
Butterhead	Hilde	25.7
	Plenos	25.9
	Borough Wonder	27.0
	Standwell	27.3
	Feltham King	28.4
	Avondefiance	28.5
	Mildura	29.8
Cos and Latin	Dorina	31.0
	Little Gem	31.0
Crisphead	Great Lakes 659	31.0
	Avoncrisp	32.8

Interaction of Light and Temperature

Blaauw-Jensen (1981) found that Pfr is inactivated at high temperature and with exposure to far-red light, but, when exposed to germination conditions, seeds with the two types of induced dormancy responded differently. A dark reaction necessary for immediate action of Pfr takes place more slowly with thermodormancy than with far-red-induced dormancy (Fig. 3.3).

Pretreatment of imbibed lettuce seeds can influence germination in subsequent light or temperature conditions. Normally, germination is inhibited by low far-red-light exposure. However, if seeds are given a short period of far-red irradiation and 1 day of dark incubation at 20°C, to establish very low levels of Pfr, and then are exposed to prechilling treatment, germination is enhanced by exposure to subsequent far-red irradiation (Vanderwoude and Toole, 1980). The lower the prechilling treatment, down to as low as 4°C, the greater the sensitivity to subsequent far-red light and the higher the germination percentage. The longer the exposure to prechilling, the greater the sensitivity. There is a decrease in sensitivity with a longer exposure

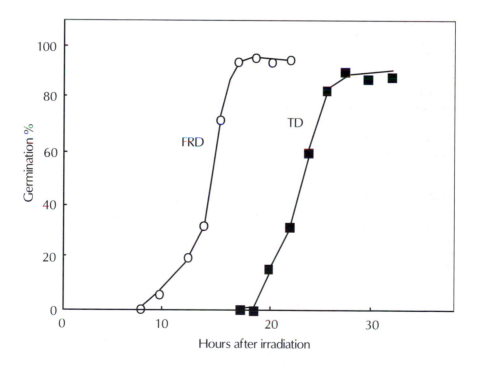

Fig. 3.3. Germination time course comparing effect on lettuce seeds of far-red dormancy (FRD) and thermal dormancy (TD). Induction by red light for 2 min. (From Blaauw-Jensen, 1981.)

to 20°C before the far-red treatment. It was suggested that the effect is due to membrane modification by the pretreatment.

Fielding *et al.* (1992) proposed that the upper temperature limit for germination was controlled by the phytochrome condition. Factors that change the temperature dependence of Pfr action will change the upper temperature limit. Treatment at 35°C, a temperature above that at which 50% of maximum germination occurs (GT_{50}), inhibited escape from far-red treatment and subsequent germination. Inhibition did not occur at 27°C.

A large proportion of the experiments on germination have been performed on seeds of the cultivar Grand Rapids, a leaf lettuce. When other cultivars were included in an experiment, variations in response dependent upon cultivar were noted. For example, Cobham Green is more sensitive to high temperature, while Great Lakes is more sensitive to absence of light (Heydecker and Joshua, 1976).

Seeds that no longer respond to red light or gibberellin are said to be in a state of skotodormancy, or secondary dormancy. However, embryos freed of seed coatings after inductive treatment will germinate normally; therefore this is a whole-seed phenomenon and not dependent upon embryonic traits (Bewley, 1980). Incubation of seeds at high temperature will induce secondary dormancy. Kristie *et al.* (1981) found that Grand Rapids lettuce possessed this trait, but New York and Great Lakes did not. However, the condition was not induced in Grand Rapids by a short high-temperature exposure, but required longer exposures, from 0.5 days to 8 days. Secondary dormancy can be prevented by repeated exposure to red light or treatment with chemicals, such as gibberellic acid (GA_3) or benzyladenine.

Several chemicals have been shown to inhibit germination. These include abscisic acid (Reynolds and Thompson, 1973) and chlormequat (Berrie and Robertson, 1973). Seed coatings also inhibit germination (see Chapter 6).

GROWTH AND DEVELOPMENT IN LETTUCE

The sequence of growth for lettuce can be divided into four stages: seedling, rosette, heading (not in all types) and reproductive. Within the seedling stage, there are three phases. First, in the germinating seed, the radicle emerges and becomes the tap root. Then the cotyledons emerge and enlarge. Lastly, the first true leaves are formed. The time to the emergence of the first true leaf is about 2 weeks.

Following the seedling stage is the rosette stage, which consists of emergence, expansion and maturation of leaves, forming either a prostrate or erect rosette. The diameter of the plant increases substantially during this stage.

Certain types of lettuce then go into the heading (hearting) stage. In butterhead and crisphead types, this consists of successive formation of cup-shaped leaves, in which earlier leaves enclose the later ones, forming a more or less spherical structure. Cos lettuces form an upright head, in which the leaves all remain erect and barely enclose each other in the longitudinal direction. Leaf formation, in both the heading and the non-heading types, continues on the compressed stem until the plants reach harvest maturity.

Harvest maturity is followed by the reproductive stage, which occurs in three phases: stem elongation, flowering and seed development. The beginning of the flowering process takes place early in the life cycle, prior to stem elongation, but the actual expansion of the inflorescence takes place during the stem-elongation process. Stem lettuces form thicker stems and stay a little longer in that process before flowering. Flower formation for all types takes place over a period of several weeks.

Emergence

After the radicle emerges, it elongates rapidly to form a tap root. Transition from radicle to primary root takes place at about 2 mm. The tap root elongates to about 3 cm after 48 h and may eventually reach a length of 60 cm or more (Jackson, 1995). Lateral root growth begins a few days after emergence. The highest proportion of laterals is formed along the upper part of the root. This differs from *Lactuca serriola*, in which a high proportion of laterals is maintained to a lower depth. This difference affects the ability of the two species to gain access to soil water (Gallardo *et al.* 1996b). Cultivated lettuce reacted to drying of the soil at 0–20 cm with reduced leaf-water status and photosynthesis, while the wild form did not. Wild lettuce has 50% of its lateral roots below 20 cm, while cultivated lettuce has only 35% below that level.

Transplanted lettuce has usually lost its tap-root apex in the seedling stage and therefore undergoes a substantial proliferation of lateral roots as the plant grows after transplanting. This effect was simulated by four methods of root pruning of 10-day-old plants of the butterhead cv. Arctic King (Biddington and Dearman, 1984). The treatments and results after 8 days were as follows:

1. Pruning of the tap-root tip had only a little effect on total root length but first increased and then decreased lateral root growth.
2. Pruning all root tips had a similar but more pronounced effect on lateral root growth. However, total root length first markedly decreased and then increased.
3. Removal of 50% of total root length, including lateral roots, severely decreased total root length and lateral root growth. This was followed by root regeneration, but neither the length of the tap root nor that of the lateral roots reached the lengths of the control.
4. The third treatment plus removal of the remaining root tips caused a severe decrease in root length, followed by recovery, but not to the control

level, while lateral growth first decreased markedly, then increased and then decreased again.

Treatments 2 and 4 also lowered shoot dry weight after 14 days, but both recovered to near control levels.

Root growth subsequent to radicle emergence can be influenced by various chemicals. Radicle development is inhibited by short-chain fatty acids and alcohols (Ulbright *et al.*, 1982) (Fig. 3.4), and by ethylene and NaCl (Huang and Khan, 1988). Kinetin inhibits and α-naphthaleneacetic acid stimulates lateral root development (MacIsaac *et al.*, 1989).

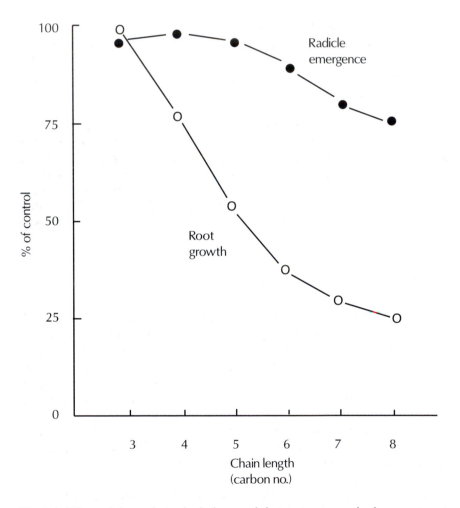

Fig. 3.4. Effect of short-chain alcohols on radicle emergence and subsequent root growth of germinating lettuce seeds. Seeds were treated in succinate-buffered alcohol solution or buffered control only. (From Ulbright *et al.*, 1982.)

Subsequent to radicle emergence, the cotyledons emerge and elongate, sustaining the plant until the true leaves emerge. Normally, cotyledons are green and unblemished. However, naturally or artificially aged seeds will develop a condition called red cotyledon or physiological necrosis, first described by Drake (1948). Small grey, brown or red lesions appear on the cotyledons. The condition is an indicator that the viability of a seed lot is decreasing. As ageing progresses, the disorder advances with increasing lesion size, followed by inability of the cotyledons to emerge. The last phase of deterioration is the failure of the radicle to emerge. Electron and light microscopy showed that the disorder is a localized deterioration, which may range from delay of the mobilization of reserves to breakdown of subcellular structure (proteins and lipids) (Smith, 1989).

To find a preventive treatment, seeds of five cultivars were tested at various temperatures and relative humidities (Bass, 1970). Storage at low temperatures, −12°C to 10°C, at various humidities, resulted in low incidence of red cotyledon for up to 210 weeks. At higher temperatures, red cotyledon became more severe.

The hypocotyl elongates as well, following radicle emergence, and is a function of both cell elongation and cell division (Galli, 1988). Cell elongation may be reduced by exposure to light. Fluorodeoxyuridine inhibits elongation of the hypocotyl in the dark by inhibiting cell elongation. Lettuce seedlings that germinated under low light intensity showed increased hypocotyl extension, although the cotyledons developed normally (Kordan, 1981). More normal hypocotyl extension could be effected by low concentrations of colchicine.

Seedling and Young Plant Growth

The seedling continues to produce true leaves, with each leaf successively broader than the one formed earlier. These leaves are produced on a shortened stem and therefore form a flattened rosette. Leaf lettuces continue to form leaves in this manner and the final mature vegetative form is still a partially erect or flattened rosette. The cos, Latin and stem types form an erect rosette, which may be partially closed at the top. Butterhead, crisphead and Batavian types form cup-shaped leaves, from which a spherical, slightly elongated or slightly flattened head develops. When the elongated condition becomes more pronounced, it is referred to as spiralling or coning. This is an unsatisfactory condition, since elongated heads are difficult to pack into cartons.

The growth rate of lettuce in early stages has been studied under controlled growth-room and greenhouse conditions. These studies would be most applicable to controlled growth production. Scaife (1973) showed there was a direct relationship between relative growth rate, based upon change in dry weight over time, and temperature. Light was held constant at a high level.

Relative growth rate was equally dependent upon net assimilation ratio and leaf area ratio in the range 10–18°C, but more dependent upon the former above 18°C. Six cultivars were grown in growth chambers. They varied in relative growth rate, but there was no temperature–cultivar interaction. The relationship between growth and temperature was sigmoidal. Field results were similar to those found in the chambers.

Verkerk and Spitters (1973) studied growth rate in a phytotron. They varied both temperature and light-energy regimes and measured the number of leaves, length and width of largest leaf and total leaf area. These traits were primarily affected by light energy and secondarily by temperature.

Aikman and Scaife (1993) developed a plant-growth model based upon time and temperature (day-degrees), photosynthetically active radiation and CO_2 concentration. Applied to growth of lettuce in a greenhouse, at four different times of the year and at four plant densities, the observed dry weight data fitted well with the predicted growth curves.

According to Bierhuizen *et al.* (1973), in greenhouses, during the period from germination to 100% soil cover by the leaves, light is more important for dry-matter production and air temperature is more important for leaf development. Light is more important for both when full soil cover by the expanding leaves is achieved. Development of soil cover is directly dependent upon the number of degree-days. They suggest that, for greenhouse production, the plants should be grown at a high temperature to reach maximum soil cover quickly. Then, when light becomes the important controlling factor, the temperature can be reduced.

Inada and Yabumoto (1989) showed that growth parameters (leaf dimensions, fresh and dry weight) were optimized at a day length of 20 h, a red to blue ratio as high as 10:1 and a red to far-red ratio of about 1:2.

The Heading Process

The use of the terms head and heart as descriptive of the arrangement of leaves, and of heading and hearting as descriptive of the process by which the arrangement is formed, can be confusing. In general, the pairs of terms are synonymous; head and heading are used in literature from the USA and many other countries, while heart and hearting are used in the UK, some European countries and Australia. Sometimes, both terms are used. In this publication, head and heading will be used.

The heading process in lettuce has been studied in the butterhead, cos and crisphead types. The earliest studies of the influence of environmental factors on heading were described in three papers by Bensink (1958, 1961, 1971). The work was in growth chambers, principally with the butterhead cultivars Meikoningin and Rapide in nutrient solutions and gravel. He studied the effect of temperature, day length and light intensity on heading characteristics, particularly the relationship of leaf dimensions to the onset of the heading process. This work applies primarily to production of winter cultivars in protected culture.

Bensink found that plants produce single leaves in succession. The rate of formation increases with increasing light intensity at a constant temperature; it also increases with increasing temperature at constant light. Leaf width responds directly to increases in day length and light intensity. Leaf length, on the other hand, increases with lower light intensity and shorter days.

The earliest leaves are long and narrow. Under high light intensity or long days, leaves become successively broader to a maximum. Under low light or short days, leaves remain relatively long and narrow. The effect of day temperature depends upon the light intensity. Leaves will increase in broadness under high light as the temperature increases and tend to remain narrow at low light intensity. Leaf-length increase, on the other hand, is slowed under high light but is faster in low light as the temperature increases. Night temperature has the opposite effect of day temperature. High night temperature encourages long narrow leaves and lower temperature increases the broadness. Cell number increases with increasing light and temperature, but cell length in the midrib is reduced, thus explaining the relative broadening at higher light-energy levels. Leaf growth at the early part of the life cycle is based on both cell division and cell growth. At later stages, growth is based upon cell-size increase.

The onset of heading is concurrent with the achievement of a leaf length/ width ratio of 0.8. However, some leaf lettuces develop broad leaves without forming heads, so the change in ratio may not be the direct cause of heading.

In cos lettuce, the principal work is by Nothmann (1976a, b, 1977a, b). Unlike butterhead lettuce, leaves of similar dimensions accumulate in the centre of the head. The tops of the leaves fold downward slightly from the outside in those cultivars which are self-folding. There is no substantial broadening of the leaves and no strong tendency of the leaves to fold inward (Fig. 3.5). Root temperatures of 12°C or 20°C were more conducive to good head formation than higher temperatures.

In crisphead lettuce, a few studies have been made. Bremer and Grana (1935) found a single, highly mutable gene, the dominant allele of which transformed head lettuce to an open cos-like rosette. Bassett (1975) found that three to five genes contributed to the heading difference between Minetto, a crisphead type, and Gallega, a Latin type with open rosette. This difference is also associated with leaf dimensions, as in the butterhead study. Wurr and Fellows (1991) studied the effect of solar radiation and temperature on head weight of the crisphead Salinas (known as Saladin in Europe). They found that head weight is positively correlated with solar radiation, but negatively correlated with temperature during the heading process.

The relative growth rate of butterhead lettuce cultivars, defined as dry-weight increase in g g^{-1} day^{-1}, was shown to increase with temperature in early

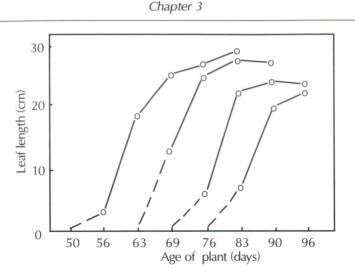

Fig. 3.5. Growth of leaves of cos lettuce. Length of every tenth leaf beginning at leaf 15. Broken line indicates that leaf was less than 1 cm at previous measurement. (From Nothmann, 1976a.)

phases of growth, but the rate declined as the plants aged (Van Holsteijn, 1980). Wheeler *et al.* (1993) found that mean relative growth rate of a butterhead cultivar grown in tunnels decreased with time in a linear fashion. There was also an interaction between time and temperature: the effect of temperature changed from a positive to a negative one. Thus the optimum temperature for relative growth rate declined from 23°C to 10°C in the period from transplanting to harvest. The application of these findings to the growing of lettuce in the field under complex conditions of temperature and day length rising and falling, but not necessarily at the same times, is not known.

Reproductive Growth

Reproductive processes are the final stages of growth in lettuce. Although the floral parts are laid down quite early in the life cycle, the actual development of the reproductive stages takes place after vegetative maturity. At that point, stem elongation commences and the stem emerges from the top of the rosette or the head. The tight heading of crisphead lettuce may hinder this development, forcing the stem to elongate in a circular manner inside the head until it can break through. This process is usually aided mechanically in commercial seed production. As the stem elongates, a terminal flower is produced, which limits the final height of the plant. Subsequently the stem branches, forming secondary and tertiary flowers. The composite flower opens about 10 days after the flower first appears. Each flower opens only once and remains open only part of the day, the period ranging from 1 h to several hours, depending upon temperature and light intensity. Fertilization takes place during the open-flower period. The achenes mature over a period of about 2 weeks from flower opening. The flowering period and the seed-maturity period for the entire inflorescence each take place during overlapping periods of 3–4 weeks.

Stem elongation is normally a response to the effect of long day length and high temperature. Long days initiate the elongation process for some cultivars, while others are more or less day-neutral. High temperature accelerates the elongation, so that flowering occurs earlier.

Various treatments can affect the onset of the reproductive process. Vernalization, or treatment of emerging seedlings with low temperature, just above the freezing-point, will cause plants to begin stem elongation early. Rappaport and Wittwer (1956) investigated the effects of: (i) length of vernalization period, (ii) stage at which vernalized, (iii) night temperature after vernalization, (iv) root temperature, and (v) photoperiod on the cultivar Great Lakes. They found that 13 days of vernalization at 5°C on imbibed seeds and on seedlings less than 3 days old were necessary to affect the reproductive process. Flowering was not accelerated at a growing temperature of 10°C, but increasingly accelerated at 15.5°C, 18°C and 21°C. Lower root temperatures delayed flowering. The combined effect of vernalization, long days and high temperature produced maximum acceleration. Low temperature also has a vernalization effect on seed developing on the mother plant (Wiebe, 1989). Exposure to temperatures of 5°C or 10°C produced higher bolting rates as compared with 15°C.

The effects of short days can also be overcome by the use of certain chemicals applied to the plant in early growth stages. The most commonly used of these is GA₃. Bukovac and Wittwer (1958) showed that Great Lakes lettuce could be induced to flower in a 9 h photoperiod, at 10–13°C, when GA₃ was sprayed at 20 µg per plant at the eight- to ten-leaf stage. For other combinations of day length up to 18 h and temperature to 21°C, GA₃ had an additive effect, increasing the percentage of plants that flowered, reducing the number of days to flowering and increasing the height of the seed stalk.

There are genetic effects on response to photoperiod (see Chapter 2). These genes can be useful in studies of the physiological aspects of bolting. Waycott and Taiz (1991) used a very early-flowering line, the double homozygote of *Ef-1* and *Ef-2*, and genetic dwarfs derived from it, to study the effects of different gibberellins on stem elongation. They found that exogenous application of gibberellins to the dwarfs would restore normal stem elongation, thus overcoming the block by each mutant of a step along the endogenous gibberellin pathway. They were also able to identify some of the gibberellin intermediates and partially define the pathway. Only one of the mutants was apparently not tied directly to gibberellin production.

Flower Differentiation and Development

Flowers are in heads, or capitula; each has 12–30 florets (Jones, 1927). Each floret is perfect and has a single ovary. The receptacle is flat and without hairs. The involucre is cup-shaped and is formed by a series of overlapping bracts. The corolla is ligulate, yellow and five-toothed at the apex. Each floret has five stamens, which are united in a tube and attached to the corolla. There is one stigma with two lobes, covered by hairs. The ovary is one-celled.

The flower parts begin to differentiate early in the growth of the stem. The oldest flower-head forms the terminal bud. The meristematic region in each flower-head is convex at first, but then becomes flat and broad. Protuberances arise over the entire surface; each will develop into an individual floret. On each protuberance, a marginal ring of cells forms, which will become the corolla. The pappus hair and stamens appear almost simultaneously, followed by the carpels.

The terminal, secondary and tertiary flower-heads together form a dense, corymbose, flat-topped panicle (Feráková, 1977). Flower opening occurs over a period of several weeks and takes place in two or three flushes of growth. Each flower-head opens only once, in the morning, and anthesis takes place during that period. It takes 12–15 days from anthesis to seed maturity, with the flushes of seed development matching those of the flower opening.

ENDIVE PHYSIOLOGY

The growth and development processes of endive that have been studied are similar to those of lettuce. However, much less work has been done with endive than with lettuce.

Germination of endive seed is inhibited by high temperature. This can be overcome by treatment with thiourea (Thompson, 1946). A higher percentage of seeds presoaked with thiourea or plain water germinated at 30°C, compared with unsoaked seeds. The water treatment was less effective than the thiourea.

Endive seeds can be primed to increase the rate and uniformity of germination of seeds under temperature stress (Bekendam *et al.*, 1987). Treatments with water or potassium nitrate (KNO_3) solution, accompanied by exposure to red light, are effective.

Seed-stalk elongation and flowering can be stimulated by treatment with gibberellin or by vernalization (Harrington *et al.*, 1957) (Table 3.2). The treatments are effective alone, and, when used together, induce elongation and flowering at an accelerated rate.

Table 3.2. Effect of gibberellin and vernalization on stem height and flowering of endive. (From Harrington *et al.*, 1957.)

Treatments	No. plants	Percentage open flowers		
		9 Oct.	26 Oct.	3 Dec.
Unvernalized				
No gibberellin	39	0	0	0
50 µg once	36	0	0	65
50 µg weekly	10	0	30	100
Vernalized				
No gibberellin	27	0	7	51
50 µg once	29	0	41	89
50 µg weekly	34	21	86	100

CHICORY PHYSIOLOGY

In addition to studies in germination and other physiological processes similar to lettuce and endive, there are two other processes in chicory that have merited study. One process includes the changes taking place in the witloof types in the transition from vegetative growth outdoors to chicon production under indoor forcing conditions. The other involves the relationships between inulin content and formation of sugars in the roots of industrial chicory.

Valette (1978) found that chicory seeds germinated best at 20–25°C. In contrast, Corbineau and Come (1990) found that the optimum was 25–30°C. Inhibition and slowing of germination occurred at low temperatures (5–12°C). Germination of light-coloured seeds, which are more mature, was higher than for dark seeds. They proposed measures to increase the quality of the seeds: (i) sorting by colour; (ii) germination testing at below-optimum temperatures; and (iii) harvesting at a suitable maturity stage. Chicory germination is inhibited in the dark (Ellis *et al.*, 1989). As dosage of light, based upon photon flux density and photoperiod, increases, germination first increases and then decreases.

Early cultivars of witloof chicory require less cold treatment before forcing than late cultivars (Huyskes, 1962). The core length in the chicon is an indicator of earliness. The longer the relative core length, as a function of the whole chicon length, the earlier the cultivar. Quality of chicons decreased with increasing core length beyond an optimum length appropriate for an early, mid- or late cultivar. Longer core length was also associated with increased proportion of reducing sugars in the root (Rutherford, 1977).

Two major root reserves supply nutrients for remobilization in the root for production of the chicon (Fouldrin and Limami, 1993). Inulin comprises 80–85% of the root dry weight and contributes carbon. Nitrogen compounds constitute about 1% of the dry weight and contribute nitrogen to the metabolism during forcing. Additional nitrogen is available from exogenous sources.

The proper balance between nitrate and phosphate availability contributes to root yield and quality and therefore to chicon yield and quality (Ameziane *et al.*, 1997). High nitrate fertilization during the vegetative period has a deleterious effect on these traits, but reduced nitrate addition reduces shoot growth and the shoot/root ratio, and leads to yield reduction. However, a moderate reduction of phosphate supplied to the plants does mitigate the effects of high nitrate, while adequate growth is maintained.

Inulin is a polyfructosan, a polymeric chain of fructose ended by a glucose molecule. It is found in Jerusalem artichoke and other *Asteraceae* as well as in chicory. Chicory roots contain 17% inulin (De Baynast and Renard, 1994). Inulin can be broken down to form a syrup, which contains mostly fructose sugar. The usefulness of chicory as an industrial source of sugar, competitive with sugar beet, potatoes and maize, depends upon increasing the total sugar yield. This can be accomplished primarily by breeding.

Chicory roots are 72–77% water. Of the dry matter, 65–85% is carbohydrate, primarily inulin, which would yield about 85–90% fructose and 10–15% glucose. Most of the rest of the dry matter consists of nitrates, minerals, lipids and sesquiterpene lactones. The sesquiterpene lactones are reponsible for the bitter taste of chicory leaves.

Although the content of total non-structural carbohydrates increased during plant growth, the content of sucrose and glucose did not (Ernst *et al.*, 1995). However, during storage, the sugars increased while total carbohydrates decreased. Another non-inulin fructan series undergoes changes during root growth and storage and may serve as an indicator of the optimum time to harvest for best chicon production (Fig. 3.6).

Witloof root explants develop differently, dependent upon whether the development takes place in continuous light or continuous dark (Demeulemeester *et al.*, 1995). Under light, flowering stems were initiated, while, in the dark, vegetative stems were formed. Treatment with various gibberellin inhibitors showed that GA_1 may be synthesized during the *in vitro* period and that it controls flower-stem growth, but not floral initiation. Treatment of chicory root explants with red and far-red light had different effects on flowering induction (Badila *et al.*, 1985). Red light enhances flowering and this effect is increased with the increase of white light as well. Phytochrome is evidently a factor in induction of the reproductive process *in vitro*.

During the root vernalization period, the free gibberellins GA_3, GA_4 and GA_9 increase in the root (Joseph *et al.*, 1983). They may augment the effect of chilling in induction of the flowering process. Cytokinins may also have a similar effect (Joseph, 1986).

Reproduction studies have been carried out in radicchio, a typical biennial-type plant. During the first year, it grows vegetatively, forming a large

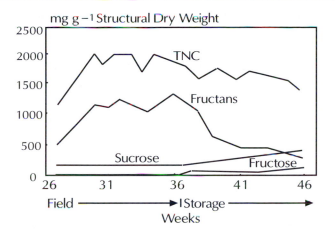

Fig. 3.6. Changes in total non-structural carbohydrates (TNC), fructans, sucrose and fructose in chicory roots through field growth and storage periods. (From Ernst *et al.*, 1995.)

tap root and a rosette of 30–70 leaves (Gianquinto and Pimpini, 1995). In the early spring of the second year, it first forms new leaves, followed by stem elongation and flowering as the day length increases and temperatures rise. The stem branches and elongates to 1.2 m and more and produces many inflorescences. However, if transplanted in late winter, bolting may be induced in the first year, either before or during head formation. Chicory has an absolute long-day requirement for flowering, but the early-cold requirement may be absolute or facultative, depending upon cultivar. Cold treatment can enhance the long-day effect on the bolting process.

The effect can be reduced by prior or subsequent exposure to high temperature. Gianquinto and Pimpini (1989) studied the influence of temperature on several aspects of spring-grown radicchio growth. Temperatures of 20–26°C were most favourable for germination and emergence, as well as for growth during the following 5–10 weeks (Fig. 3.7). This temperature range was also important in providing a devernalization effect, thus prolonging vegetative growth. The delay in bolting resulted in increased yield.

PHYSIOLOGICAL DISORDERS

Lettuce

Lettuce is affected by a number of physiological disorders, i.e. disorders in which no organism or virus-like entity has been identified as a cause. Rather,

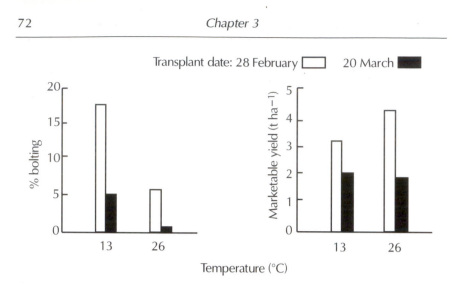

Fig. 3.7. Effect of temperature during emergence and of transplant date on percentage of bolting at 113 days and marketable yield at 132 days of chicory cv. Rosso di Chioggia. (From Gianquinto and Pimpini, 1989.)

the disorders have been ascribed to environmental causes, such as temperature, moisture, storage conditions and other stresses.

Tipburn

Tipburn is a disorder that usually occurs at the time of harvest. Therefore, damage due to tipburn may result in the loss of a field of lettuce and all the expenses of growing the field to maturity. It can occur on all forms of lettuce. Symptom expression begins with the appearance of small brown spots close to the margins of a leaf. These may also be accompanied by necrosis in small veins in the area. As the disorder progresses, the necrotic areas coalesce, forming a lesion, which may be quite small, or may be several centimetres in length along the margin and 1–2 cm wide (Fig. 3.8). In crisphead lettuce, tipburn usually occurs within a day or two of market maturity and on leaves just inside the circumference of the head to several layers deep. Sometimes symptoms are visible on the outside leaves. Plants grown in growth chambers or greenhouses may show tipburn symptoms on exposed leaves at a relatively early stage of rosette growth. Damage on butterhead lettuce is usually more extensive than in crispheads, and usually shows on the exposed leaf margins at the top of the head as well as in the interior. Cos lettuce shows lesions at the tips of partially exposed leaves. Leaf lettuce develops symptoms on margins of partially exposed middle leaves.

Thompson (1926) reviewed the tipburn literature from 1891 to 1925 and described the causes of tipburn as proposed by the early authors. He evaluated some of the earlier proposals, and concluded that carbohydrate content, irrigation and other factors affecting growth were involved in

Fig. 3.8. Symptoms of tipburn on internal head leaves of crisphead lettuce.

causing the disorder. Andersen (1946) specified increasing difference between soil and air temperature as leading to tipburn. Kruger (1966) first implicated calcium metabolism as a part of the tipburn development progression. Following that discovery, Thibodeau and Minotti (1969) were able to control tipburn on a butterhead lettuce cultivar, Meikoningen, with foliar sprays of calcium salts. They were also able to accelerate the development and increase the severity of tipburn with application of oxalate, which ties up soluble calcium in tissue. They suggested that calcium distribution in the plant and not just availability from the soil dictated whether tipburn would occur. Any factor or combination of factors limiting the supply of calcium to inner leaves in relation to the growth rate of those leaves will induce tipburn. Application of foliar calcium directly to expanding inner leaves is not possible with normally heading crisphead lettuce at late stages of development and this procedure does not work with this type of lettuce (Misaghi *et al.*, 1981b).

The events that cause the shortage of calcium in marginal tissue apparently start with an accelerated growth rate. The causes that earlier authors attributed as single causes of tipburn have in common a similar acceleration effect. These causes are: an increase of temperature, an increase in light intensity, an addition of nitrogen fertilizer, an irrigation application or other growth stimulant. These act, singly or in combination, to stimulate an increase in growth rate in marginal tissues of, usually, expanding inner leaves. The key to the influence of calcium is that it translocates slowly in the transpiration stream, and therefore fails to keep pace with the expanding marginal tissues. Subsequent events are the expression of tipburn symptoms.

It is not known what the exact contribution of the calcium shortage is. However, it has been shown in a number of studies that calcium is important to cell-wall strength and membrane integrity. Damage to these structures leads to cell collapse, discharge of latex from the laticifer system and the discoloration associated with the disorder. Other factors may be involved. See Collier and Tibbitts (1982) for an extensive review of the literature and a model for tipburn development, and Shear (1975) for a discussion of other calcium-related disorders.

The development of tipburn results from the combination of adverse environmental conditions, changes in chemical pathways, changes in growth characteristics and relative susceptibility of a plant to these influences. Each of these contributing causes can be modified to attempt to achieve control of the disorder.

Environmental causes, such as light and temperature, cannot be controlled for outdoor field plantings of lettuce, but can be partially controlled in greenhouses and largely controlled in growing rooms. In the latter cases, light and temperature can be adjusted to avoid sudden changes in growth rate. In all environments, application of water and nitrogen can also be controlled to alleviate sudden increases in growth rate. This might include the use of trickle irrigation instead of other massive, intermittent applications of water (Cox and Dearman, 1981).

Plants deficient in boron may develop tipburn symptoms, which suggests that the addition of boron may contribute to the prevention of tipburn (Crisp *et al.*, 1976).

A number of studies have shown that there is substantial variation in the ability of various cultivars to resist tipburn, even when the environmental conditions are favourable for its onset. Cox and McKee (1976) tested eight cultivars in seven field plantings over 2 years. They found that four cultivars, Borough Wonder, Cobham Green, Lobjoits Green Cos and Avoncrisp, consistently showed severe tipburn symptoms. Four other cultivars, Avondefiance, Great Lakes 659, Little Gem and Webbs Wonderful, were virtually free of the disorder. This indicates that there are consistent differences in tipburn reaction among cultivars, and these differences are not necessarily related to type. However, the same cultivars were grown in the greenhouse and results were different from those in the field. Only Great Lakes 659 was free of symptoms, while Avondefiance, Little Gem and Webbs Wonderful were severely affected. Borough Wonder was susceptible in both environments, but Cobham Green and Lobjoits were moderately susceptible in the greenhouse, and Avoncrisp was resistant in the greenhouse.

Screening for tipburn resistance among cultivars can be accomplished either in the field or in more controlled conditions. In the field, several trials should be grown, since the conditions for induction of tipburn cannot be predicted. Misaghi *et al.* (1981a) suggested a method for testing under more artificial conditions. They induced tipburn in detached mature heads grown at

30°C and showed differences among cultivars similar to those found in the field. Salinas, Calmar and Montemar all showed resistance in both environments, while Monterey and Calicel showed susceptibility in both. Greenhouse results may or may not be consistent with field observations, as suggested by Cox and McKee (1976).

Rib disorders

Rib blight, known also as rib discoloration and brown rib, is the occurrence of yellow, brown or black streaks along the midrib and secondary ribs of cap leaves and of leaves just below the cap leaf. These occur on mature heads of crisphead lettuce just before harvest; the disorder has also been anecdotally reported in other types of lettuce. The symptoms of the disorder do not materially change after harvest, but may serve as sites for secondary infections. There is no known cause, although it seems to be associated with high day and night temperatures. Jenkins (1962) described moderate resistance in a breeding line of crisphead lettuce, but no breeding for resistance has been reported.

Pink rib is a field disorder of crisphead lettuce (Marlatt and Stewart, 1956). It may occur on other types of lettuce but has not been reported. Pink discoloration may appear on the main rib as the lettuce reaches maturity, although it is more commonly found in overmature lettuce. No organism has been unequivocally associated with the disorder. It may be accelerated in development at higher than normal storage or transit temperatures. There are differences among cultivars in the development of the disorder, but no genetic or breeding research has been reported.

Atmospheric gas effects

A number of atmospheric gases have been shown to have toxic effects on lettuce. These include ozone, sulphur dioxide, nitrogen dioxide and peroxyacetyl nitrate (PAN). Lettuce is affected only sporadically except where it is grown near pollution sources. Plant damage can include visible discoloration, pitting or necrosis of leaves, which on lettuce can render the crop unsaleable. Also, relatively lower levels of these materials may not cause visible cosmetic damage, but may affect growth and yield. Little work has been done on variation in degree of susceptibility or breeding for resistance. Reinert *et al.* (1972) showed differences in sensitivity among eight lettuce cultivars.

Postharvest disorders

Most diseases and disorders usually originate before harvest. During the postharvest period, the effects of tipburn, pink rib, rib blight and various viral, fungal and bacterial diseases may be accentuated by further deterioration caused by secondary organisms. Several problems often originate after harvest, during handling, storage, transportation or marketing. The most important of these are russet spotting, brown stain, bacterial soft rot and grey mould. The latter two are usually secondary effects, following previous field-originated disorders, diseases or mechanical injury.

Although it may occasionally appear in the field, russet spotting is primarily a postharvest disorder. Russet spotting occurs when lettuce is exposed to ethylene gas. Ethylene is produced by fossil-fuel-driven machinery in the cooling and storage areas, and by several kinds of ripening fruit, such as tomato or melon, either in the transported load, on the market shelf or in the refrigerator. Levels as low as 0.1 p.p.m. can cause the disorder on lettuce (Moline and Lipton, 1987). Other factors influencing the development of the disorder include temperature, O_2 and CO_2 levels, maturity and cultivar. Ke and Saltveit (1986) found that calcium and 2,4-dichlorophenoxyacetic acid (2,4-D) inhibited russet spotting. Russet spotting is expressed as small, sunken, rust-brown spots, which appear on ribs of outer head leaves, and may progress to inner leaves as well (Fig. 3.9). When severe, the spots may coalesce. The disorder may be controlled by use of machinery and vehicles in the cooling and storage areas that are not powered by fossil fuels. It can also be prevented by forbearing from shipping, storing or refrigerating lettuce with fruits that produce ethylene. There are differences among cultivars in susceptibility to damage, but there are no active breeding programmes for resistance.

Brown stain is a disorder caused by excess CO_2. The gas is a normal product of respiration, but may reach higher than normal levels when there is little or no gas exchange, as in storage rooms, freight carriages or lorry

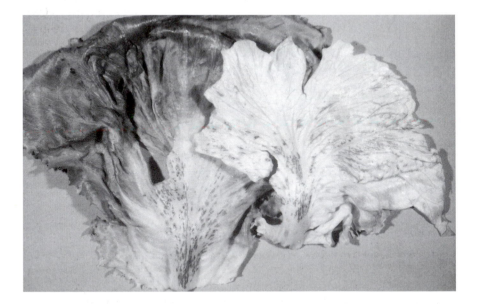

Fig. 3.9. Russet spotting symptoms on lower leaf blade and midrib of crisphead lettuce.

trailers. The effect is expressed in two ways. Brown lesions appear about the midrib of enclosed leaves, but not leaves closest to the interior of the head. These are usually about 5 mm × 12 mm, but may be smaller or may coalesce to form larger lesions. If the lesions are small they may be confused with russet spotting. However, excess CO_2 may also cause a reddening of the smallest interior leaves. Brown-stain expression may be increased at a lowered O_2 level or by increased carbon monoxide (CO) as well as high CO_2 (Kader *et al.*, 1973). This may occur when lettuce is kept under modified atmospheres to reduce discoloration. Brown stain may be controlled by maintaining an adequate O_2 level, a low CO_2 level and/or low CO. There may be sensitivity differences to CO_2 among cultivars, but little research has been done.

Bacterial soft rot occurs on all types of lettuce, as a secondary effect on tissue that has been invaded by fungi or is physically damaged or dead. It is one of the two most serious market diseases of lettuce (Ceponis, 1970). Soft rot can be caused by one or more of the following: *Pseudomonas marginalis* (Brown) Stevens, *Pseudomonas cichorii* (Swingle) Stapp and *Erwinia carotovora* (Jones) Bergey (Moline and Lipton, 1987). It originates as small brown spots similar to those of russet spotting, but the veins may become discoloured as well. In severe cases the whole head may become slimy. The disease may be minimized by rapid cooling, holding lettuce at the proper storage temperature and shipping lettuce with no disease or mechanical damage.

The second major market disease of lettuce is grey mould, which is caused by *Botrytis cinerea* Pers ex Fr.(Moline and Lipton, 1987) (Fig. 3.10). The organism also causes disease problems in field and greenhouse plantings (see fungal diseases, Chapter 7). Tissue affected by grey mould appears water-soaked and may range in colour from grey to green to brown. At late stages of development, the tissues become soft and slimy and the distinctive smoky grey mycelium appears. Although the organism grows at a wide range of temperatures, rapid cooling to the range 0–2.5°C minimizes the damage. Control is also possible by taking precautions to reduce physical injury. Elia and Piglionica (1964) found evidence of resistance in some cultivars, but no breeding work has been done.

Endive

Brownheart is a disorder that affects endive and escarole (Moline and Lipton, 1987). It is manifested as marginal browning on immature leaves within the plant. There is no organism associated with the disorder. Rather, it appears to to be similar to tipburn of lettuce in that it is also related to poor calcium distribution in rapidly growing tissues. However, since endive types form relatively open heads, it is possible to control the disorder with a foliar spray of a calcium compound (Maynard *et al.*, 1962). Treatment with calcium salt solution twice weekly reduced the incidence of brownheart about eightfold.

Fig. 3.10. Smoky grey mycelium of *Botrytis* on leaf of crisphead lettuce.

Heads with brownheart may be subject to bacterial soft rot as a secondary problem during the marketing process. Maintaining the temperature close to 0°C can minimize the rate of decay.

Chicory

Krahnstover *et al.* (1997) described 20 physiological disorders of witloof chicory that may occur during the forcing process. They suggest that preventive measures during root production, storage and harvest and during all postharvest phases are extremely important (see Chapter 4). The disorders include: loose, unfilled or open heads; rosetted, shortened heads; black-blue leaf tint; blind root (axial buds with no main bud); brown flecks on the central axis; brown leaf edges; low temperature damage (oval red-brown areas on outside leaf surfaces) and black (necrotic) flecks on leaves.

Internal browning, a discoloration of the central part of the stem, was described by Den Outer (1989). It appears to be a calcium-related disorder, similar to lettuce tipburn.

PRODUCTION METHODS

LETTUCE

Lettuce in its various forms is grown nearly everywhere in home gardens. One can usually choose a type to grow that is adapted reasonably well to an environment and a season. However, for commercial production, lettuce is somewhat restricted geographically and seasonally. The growing of a high-quality product is dependent upon suitable climate and soil.

Lettuce is a cool-season crop that requires good soil and an adequate supply of water. In areas that are naturally cool most of the time, such as coastal valleys in the western USA, lettuce can be grown nearly all year round.

Lettuce does best when the daytime temperatures are from 18°C to 25°C and the night temperature from 10°C to 15°C. The coastal districts of California in the summer, the deserts and Florida in winter, south-eastern UK in summer and the Mediterranean coast of Spain in winter are examples of lettuce-growing locations with a suitable combination of climate and season. Other areas, such as New York and continental Europe, are warm in the summer and require cultivars selected for heat tolerance.

Lettuce is grown on a variety of mineral soils, from those containing much clay to those with substantial sand. Clay soils are heavy in texture, hold water well and tend to be cool in summer and cold in winter. Sandy soils are light, do not hold water well and are warmer than clay soils. Lettuce planted in sandy soils is more prone to tipburn than lettuce planted in heavy soils. Lettuce is also grown on organic soils in certain areas, such as the north-east, the Midwest and Florida in the USA, which are usually referred to as muck or peat soils. These are created from drained marshlands. They consist largely of lightweight particles of organic matter. These particles are easily wind-blown and oxidized and the soils consequently have a finite life.

Lettuce grows best on fertile soils that are slightly acid to mildly alkaline (pH 6.5–7.2). Highly acid organic soils require the addition of liming materials. Fertility levels should be adequate for nitrogen (N), phosphorus (P) and potassium (K) as well as for a number of the minor elements. A study by Costigan (1986) showed that differences in soil types from different areas can have a marked effect on early growth of lettuce. The soils were set in miniplots at a single location to eliminate the effects of climate and cultural practices. Growth differences appeared early and were related primarily to the availability of adequate phosphorus.

Lettuce is moderately sensitive to high salinity in the soil. High salinity is common on arid lands, especially desert soils, as in California and Arizona. Salinity can prevent germination; if germination takes place, subsequent growth is slow and the plants are stunted. There are several methods of dealing with salinity.

One is to plant a more tolerant crop species. Another is to flood the land to leach the salts below the root zone. A third is to use tolerant cultivars. Shannon *et al.* (1983) tested 85 cultivars and found substantial differences among them. Climax, Climax 84, Shawnee, Tom Thumb, Fulton and Wintergreen were the most tolerant. In a subsequent test, Shannon and McCreight (1984) found a wider range of variation among plant introductions of *L. sativa*. Pasternak *et al.* (1986) tested crisphead and cos cultivars under a range of salinity levels and found that the cos types were the most tolerant.

Lettuce requires a substantial amount of water for good growth. Most lettuce-growing areas use some form of irrigation for the entire water requirement, or as a supplementary source in some areas where natural rainfall supplies most of the water. As lettuce does not perform well under saline conditions, irrigation water should be relatively salt-free.

Land Preparation

Soil for lettuce should be friable and clod-free. This provides a better seed-bed for direct-seeded lettuce and a better planting bed for transplanted lettuce. In most commercial production a series of implements are drawn by tractor through the field to bring this about. First the preceding crop is disc-harrowed to begin the process of breakdown of the stems, leaves, roots, etc. in the soil. The field may be ploughed to turn under the debris and then disc-harrowed at intervals to complete the breakdown and smoothing process. The field may also be disc-harrowed several times without ploughing. If the lettuce is to be grown on raised beds, the beds are formed (listed) following application of preplant fertilizer. Many fields may have herbicide applied before planting. The field may be irrigated before planting to facilitate tillage. In some of the eastern US lettuce-growing areas and in Europe most lettuce is grown on flat ground and preplant materials are applied without listing.

At intervals of several years, fields may be levelled with a landplane. Modern planing is done with a laser beam to guide the plane vertically so that a very high degree of levelling is obtained. Also, at intervals, the land may be subsoiled (ripped) to break up a hardpan and allow water to filter freely through the soil. Soil compaction of mineral soils can be a problem on lettuce land because heavy equipment, including tractors, trailers and lorries, pass over the field at harvest. The soil may be quite moist at this time, which would exacerbate the compaction problem.

Conventional tillage practices between crops tend to compact the soil and reduce the penetration of roots and water through the soil. Minimum tillage can alleviate the problem; two procedures may be used: surface minimum tillage and deep minimum tillage. Surface tillage is designed to till only the top few inches of soil and is particularly useful when subsurface drip irrigation is practised. Deep minimum tillage is used in sprinkler- and furrow-irrigated fields to reduce soil compaction, increase aeration and soil tilth and reduce the amount of time between crops.

Other treatments that may be used in land preparation include fumigation under plastic to reduce the population of undesirable soil organisms, and flooding to reduce salinity.

Planting

There are two principal means of planting outdoor lettuce: direct drilling (seeding) or transplanting seedlings. The use of these methods for lettuce is regional. Nearly all lettuce grown in the USA is direct-seeded. In Europe and other places, a very high percentage is transplanted.

The method of direct seeding of lettuce has changed considerably in recent years. At one time, the seed was sown, uncoated, by a planter much like a grain drill. As many as 60 seeds m^{-1} were sown, amounting to over 1 kg ha^{-1}. Now most seed is sown by use of a space planter with holes at intervals on a belt, which drops each seed singly, and spaced at a specified distance, usually from 5 to 10 cm. Nearly all seed is pelleted, with coatings of different materials, to provide a more rounded unit, which is more easily handled by the equipment than the highly angular uncoated seed (Fig. 4.1). Lettuce seed is sown at relatively shallow depths, ranging from just over 1 cm to virtually at the surface. The latter practice is used for plantings made under high temperature conditions, such as in desert locations in the late summer or early autumn.

In the western USA and in many other lettuce-growing areas, seed is sown in raised beds, usually with two rows per bed (Fig. 4.2). The rows are usually 36 cm apart and the bed centres are usually about 1 m apart. The beds are usually raised about 10–15 cm. In other locations in the USA and elsewhere, seeds are sown in single rows on flat ground; the rows are usually about 40 cm apart. Low beds with multiple rows are also occasionally used.

Bed direction may be in either east–west or north–south orientation. In

Fig. 4.1. Coated lettuce seed (left) and uncoated seed (right).

Fig. 4.2. Typical planting of crisphead lettuce on two-row raised beds, Salinas Valley, California.

Salinas Valley plantings, the side with either the southerly or easterly exposure was the warmest (Zink, 1967). Growth in the east row was superior to that in the west row, but in the other orientation the south row was only superior to the north row in the early spring plantings, when temperature was more critical.

Immediately after seeding, a pre-emergence herbicide may be sprayed over the row if a preseeding material has not been applied. Then the field is usually sprinkler-irrigated once a day for several days to germinate the seeds and start emergence. In some cases, the seed is sown in moisture and no germination sprinkling is necessary. In hot climates, the sprinklers are turned on at the end of the first day to cool the seed and the soil and to encourage the onset of germination during the cooler night hours.

The seedlings are allowed to grow for 3–4 weeks, at which time they are thinned to the desired field spacing, about 25–30 cm. In the USA, thinning is usually done with a long-handled hoe. In earlier years, a short-handled hoe was used, but this has been banned in a number of places because it requires bending over and can cause back injury to the worker.

Little transplanting is done in the USA, while in the European lettuce-growing areas this practice is standard. Lettuce seed is sown under protective cover. Single seeds are placed in peat blocks, and seedlings are grown to the three- to four-leaf stage, which takes about 1 month (Fig. 4.3). These are then taken to the field and placed in the soil at the desired spacing. When practised in the USA, single seeds are placed in cells of trays containing a planting mixture. Transplanting can be done by hand, by a semiautomatic planter or by

Fig. 4.3. Lettuce transplanted in peat blocks on soil surface, near Versailles, France.

a completely automatic device. The advantage of transplanting is that the plants are in the field for a shorter period of time and can be placed in the field immediately at the desired spacing for growing to maturity. For crisphead lettuce in the USA, transplanting has been found to have two disadvantages. One is that the tap root is pruned early and the subsequent root growth is shallower than normal. The other is that it is expensive. Eastwood and Gray (1976) found that transplanted crops in the UK are more profitable than direct-seeded crops when the price of lettuce is high, because transplanted lettuce is more uniform and the harvest density is greater. The advantage is lost when the price is low.

A second method of seeding used in some areas is called fluid drilling (Currah *et al.*, 1974). Seed is pregerminated and then sown in a liquid gel medium. Germination is faster and more uniform and reaches a higher percentage than with dry seeds. Two disadvantages are: (i) seed cannot be uniformly spaced in the row; and (ii) precise timing is very important to avoid losses due to drying or seed damage.

Vacuum planters are occasionally used for seeding. These have rotating needle-like tubes, in which a vacuum is drawn to pick one seed per tube and released to drop the seed into the row.

Cultural Practices

These include irrigation, fertilization and cultivation, as well as weed, insect and disease control. In those areas where irrigation is an integral part of the production of lettuce, sprinkler irrigation is practised at least for the early part of the growing period and then followed, some time before the heading process begins, by irrigation through furrows on either side of the bed (Fig. 4.4). In some cases, particularly if the field is not level, sprinkling is continued right up to harvest. This practice, however, can sometimes lead to disease problems. A relatively new practice with lettuce is the use of surface or subsurface drip irrigation. This minimizes the use of water by placing it only in the root zone. Usually the first irrigation is with sprinklers to promote germination and emergence, and the remainder of the water for the growing cycle is applied in the drip system.

Most sprinkling is done through nozzles attached to pipe laid in the rows. Sometimes, water is applied with a rotary boom, which moves in a large circle, or with a linear rig on wheels, which moves the length of the field.

The number of water applications on irrigated land varies with the texture of the soil, the length of the growing season and the temperatures during the growing season. There may be as few as two irrigations in cool, heavy soil during cool weather, or it may be done as frequently as weekly in warm soils during relatively high-temperature periods.

Peat or muck plantings are often on ground where the water-table is close

Fig. 4.4. Lettuce field under sprinkler irrigation, Imperial Valley, California.

to the surface. Water is in ditches surrounding the field and the field may be irrigated by stopping the water flow in the ditches to raise the level of the water-table, thus watering the plants from below.

Water-management studies have become intensified in recent years because water availability has become, or has the potential to become, a limiting factor in many lettuce-production areas. This is especially true in arid climates, as in the western USA and Spain. Gallardo *et al.* (1996a) found that evaporation was the major component of water loss.

Nitrogen and water interactions can be particularly important in drip systems. Thompson and Doerge (1996) found that a soil water-tension level of about 7.0 kPa and applied N at about 245 kg ha^{-1} are optimum for lettuce in desert conditions.

In most modern plantings of lettuce, chemical fertilizers are used, in various forms, to provide nutrients. In certain organic farms, only manures and cover crops are used. Manures are the principal form of fertilizer in less developed farming situations.

Chemical fertilizers are usually applied before seeding and two to three times during the growing period. A preplant fertilizer application may contain all three major elements or may contain phosphorus alone. Phosphorus is essential for early growth. In some soils, potassium is in good supply and is rarely added. Nitrogen becomes increasingly important through the growing process and may be added after thinning and several weeks before harvest. Minor elements are usually present in sufficient amounts in N–P–K-type fertilizers and supplementation is usually unnecessary. Organic soils are usually short of phosphorus, potassium and copper and these must be added. Calcium is added if liming is required on highly acid soils; it may be a component of a nitrogen fertilizer.

Many fertilizer trials on different types of soils under various environments have been conducted and recommendations made based on the results
of these trials. These have usually been specific for the time and place.
Relatively few studies have been made that have more general application.
Zink and Yamaguchi (1962) studied the growth rate and nutrient absorption
of the crisphead lettuce cultivar Great Lakes, from which they derived a
number of principles. The nutrient uptake curves of nitrogen, phosphorus,
potassium, calcium, magnesium and sodium all indicated that 70% of their
uptake occurred during the last 21 days before first harvest (Fig. 4.5). This
corresponds to the growth rate of the plant (Fig. 4.6). Growth rate is based
directly upon the rate of leaf-area increase.

Their conclusions particularly apply to the application of nitrogen.
Nitrogen is easily leached and early application of a large amount is wasted
since the major growth requirement is late in the growing period and much of
the early nitrogen would be lost as soil water moves out of the root zone. They
recommend that about a third should be applied at thinning and another third
about 1 month before harvest.

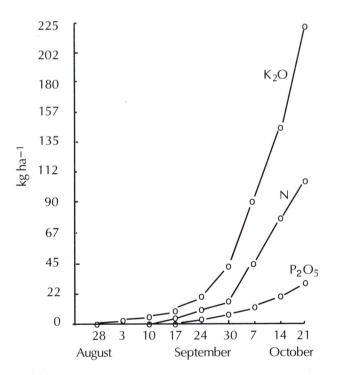

Fig. 4.5. Uptake of nitrogen, phosphorus and potassium by lettuce plants during
growth cycle, expressed as N, P_2O_5 and K_2O. Sown 5 August. (From Zink and
Yamaguchi, 1962.)

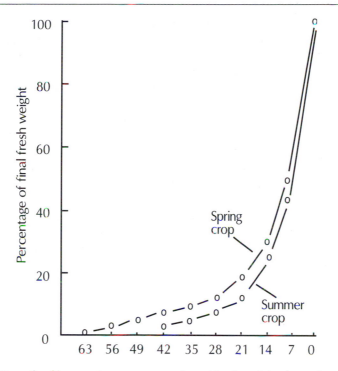

Fig. 4.6. Growth of lettuce, in percentage of total fresh weight, for spring and summer lettuce. (From Zink and Yamaguchi, 1962.)

In addition to nitrogen, for optimum growth and development lettuce requires certain amounts of other elements. These include the other major elements, phosphorus, potassium and calcium, and at least nine minor elements (Marlatt, 1974).

Mineral Deficiencies and Toxicities

Deficiencies and toxicities are caused by improper fertilization practices, unusually low or high levels of mineral nutrients in the soil or factors that limit the availability of these nutrients to the plant. For information on nutrient effects, see Marlatt (1974) and Davis *et al.* (1997).

Symptoms similar to corky root can be caused by nitrogen toxicity (van Bruggen *et al.* 1990a). Symptoms are similar but not identical to those caused by *Rhizomonas suberifaciens*. Nitrogen toxic effects appear as reddish-brown lesions, while the bacterial lesions are greenish brown. Also the corkiness caused by the toxicity is not as severe. In the field, both types can appear simultaneously, and high nitrogen levels in the soil can also increase the damage by the organism.

Other Environmental Factors

Lettuce is grown on a wide variety of soils and its tolerance to pH differences is also relatively wide. The optimum pH is in the range 6.5–7.2 on mineral soils and slightly lower on muck or organic soils. A highly acid soil, below 5.2–5.5, is deleterious to lettuce, primarily because of aluminium and iron toxicity, as well as calcium, phosphorus, magnesium and molybdenum deficiency. The effect of these mineral availabilities can be alleviated by the addition of various liming materials to increase the pH. Lettuce can be grown in alkaline soils, up to pH 8, but the addition of an acid to the soil will temporarily bring the pH down to a more optimum level.

Lettuce is moderately sensitive to soil salinity. Plants are affected by chloride and sulphate salts of sodium, calcium and magnesium. In arid areas, where salinity is likely to occur, lettuce production is restricted to those locations in which the salinity is low or can be modified. Germinating seedlings are considerably more susceptible than more mature plants and may be killed outright. Plants that survive to maturity are likely to suffer a decrease in weight and yield. Excess salinity will cause puffiness, thick leathery outer leaves and bitterness of surviving plants. Acceptable salinity levels can be achieved by flooding before planting to leach salts below the root zone of young seedlings. While the plants are growing, sprinkler irrigation will continue to keep the salts down. If the grower then switches to furrow irrigation, salt will move above the root zone and to the centre of a two-row bed. The salt will be deposited as the water evaporates, and the roots will be protected. A single row in the middle of the bed may well be damaged. There are differences among cultivars in tolerance to salinity (see above).

Applied materials, such as herbicides and insecticides, can cause plant damage, particularly when not used according to label. Ethylene, emitted from various natural and man-made sources, may be damaging, specifically by causing russet spotting.

Cultivation practices vary. Cultivation between rows and/or beds is practised at intervals during the growing season. This may be done immediately after thinning and perhaps three to four times thereafter. The purposes of cultivation are to remove weeds and help water penetration.

Weed control in the early part of the growth period is usually obtained with an application of herbicide immediately before or after planting. Subsequently, those weeds that are not eliminated chemically are hoed out at thinning and may also be hoed a second time before harvest.

Diseases and insects are controlled in one or more of several ways. Chemical control is most widely practised. Materials are applied at appropriate times for various insect pests and diseases. Some diseases and insects are controlled genetically by host-plant resistance. Certain soil-borne diseases can be at least partly controlled by planting in light soils at certain periods and heavier soils at other times. It is now considered most desirable to control pests by using integrated pest management (IPM) practices. These may include host-plant resistance, chemicals, pheromone application for mating disruption, trap crops, elimination of alternative hosts and use of parasites or predators (see Chapters 7 and 8).

ENDIVE

Endive has soil and climate requirements similar to lettuce. It thrives in relatively cool weather, although it has a somewhat higher tolerance for high temperature, high humidity and rain than lettuce. In the western USA it is grown on two-row raised beds with similar dimensions and spacing to those for lettuce. Endive is usually one of several leafy vegetables grown in adjoining narrow strips in the same field. The mixture usually includes three or four types of non-heading lettuces, endive and escarole and often the Asian leafy crucifers bok choi and Chinese cabbage. Land preparation, planting procedures, cultivation practices and control of weeds, diseases and insects are the same as for lettuce.

In other areas, endive is grown either on beds or on flat single rows, according to practices used for lettuce. Cultivation and protective practices are very similar to those of lettuce.

CHICORY

Chicory-production practices, and the cultivars used, vary according to the purpose for which it is grown. When chicory is grown as an annual for the fresh green or red leaves, it is produced in essentially the same way as endive and lettuce. In Europe, the land preparation, seeding, transplanting, irrigation and cultivation practices would be similar to those of endive and lettuce as grown in Italy and France.

Because of the possibility of cold-induced bolting, radicchio types are planted in late spring or summer. The grower may plant six or seven cultivars, scheduled sequentially according to different cold requirements for head formation. These will mature at intervals through the autumn and early winter.

A different set of cultivars is grown for root production for roasting. The leaves of these cultivars are considered too coarse for fresh consumption.

Some roasting-type cultivars have been subject to selection for adaptation to industrial use. New cultivars have also been developed that have high carbohydrate content, in the form of inulin, which is used for production of sugars. The method of production is similar to that of sugar beets. Deep, light silty or sandy, well-fertilized and well-watered soils are important to produce large, smooth, unbranched roots. These chicories are normally grown once in a 5-year rotation. Preceding crops should not be sugar beet, maize, potato or oil-seed rape, which can cause weed or herbicide residue problems for the chicory crop. The seed-bed should be both fine and firm. Seeding is usually done between mid-April and mid-May for a long growing season to achieve greater yield. However, too early seeding when the temperature is low may induce premature bolting. Seeds are pelleted and sown about 10 cm apart and at a depth not exceeding 1 cm. Precision sowing favours uniform root size.

The most highly specialized production methods are for the witloof chicory cultivars. These are grown as biennials. The first, vegetative phase of growth is outdoors, and is designed to produce a rosette of green foliage and large, uniform, single tap roots. The crop is grown in single rows or narrow bands on flat, deeply tilled, mineral soils or on double-row ridges, similarly to carrots or parsnips. Sowing is done in the late spring, after 1 May, for uniform emergence and to avoid cold-induced premature bolting. Nitrogen fertilization should be minimal to avoid luxuriant top growth and too small roots. The crop is grown for 70–100 days. Early cultivars are harvested in August and September, mid-season cultivars in October and early November and late cultivars after mid-November. The plants are mowed several centimetres above the crown and the roots are lifted. The roots are cleaned and trimmed to a uniform length and stored at low temperature and high humidity to reduce respiration and maintain turgor and for vernalization. They are removed from storage for forcing after a few weeks to a few months, depending upon the earliness characteristic of each cultivar and the forcing production schedule.

The second, forcing stage is reproductive, but does not progress beyond slight stem elongation. Forcing of the roots may be done in several ways. The oldest method is to plant the roots upright and side by side in trenches, with fermenting manure, electric cables or air ducts underneath to provide heat. The tops are then covered with about 20 cm of soil or sand, irrigated and allowed to produce the chicons. Forcing may also be done in a barn or shed, and the tops may be covered with soil or left without cover. In either case, the chicons must grow in complete darkness. Soil cover was required for earlier cultivars, which did not form compact chicons otherwise.

A new forcing procedure has essentially replaced the older methods. In this method, which was developed in the 1960s, the roots are forced hydroponically, in the dark and without soil cover (Fig. 4.7). Roots are placed upright and side by side in trays about 15 cm deep and with dimensions to give about 1.5 m^2 of surface. Trays are stacked and can drain from the top tray to the bottom. The roots are fed hydroponically with nutrient solution to promote growth of secondary roots and the growth and expansion of the apical bud to form the chicon. The trays are kept in the dark in a suitable building, although a yellow-green light may be used to allow periodic inspection. The forcing is carried out at an air temperature of about 15°C. Higher temperature causes rapid growth of loose, elongated chicons, while lower temperature slows

Fig. 4.7. Forcing witloof chicory in hydroponic growth room.

growth, producing shorter, tighter chicons. The solution is maintained about 2–3°C above air temperature to promote optimum growth. The forcing period may be as short as 20 days or extend to 30 days. Because the temperature is well controlled, forcing and harvesting schedules are more predictable than with the older forcing methods.

Krahnstover *et al.* (1997) suggest a number of preventive procedures and treatments during the entire preforcing and forcing sequences to avoid the physiological anomalies discussed by them (see Chapter 3). These include proper density of seed in the field (360,000–440,000 seeds ha^{-1}) and adequate nitrogen and potassium fertilization to produce roots large enough to support growth of the chicon. During the storage period, it is important to maintain relative humidity at 98% and temperature at −1°C. Regulation of the nutrient solution pH and of the carbon dioxide (CO_2) content of the air during the forcing period are also important.

MESCLUN

Mesclun is a French word meaning mixed and refers to a salad mixture of mostly immature leaves of various species (Fig. 4.8). Mesclun apparently originated in the Piedmont and Provence areas of Italy and France, respectively. Mesclun may include combinations of the following: red and green romaine, frilled red leaf, red and green oak-leaf lettuces, plus endive,

Fig. 4.8. Variation in shape and appearance in mesclun or spring mix.

radicchio, arugula, spinach, beet tops, red Swiss chard, mizuna and tatsoi. The various components are usually direct-seeded separately, in multiple rows (usually six) on low beds. The lettuce components are sometimes seeded as a mixture. The seeds may be germinated under plastic in cool weather. Irrigation is by sprinkler.

PROTECTED GROWTH

The growing of lettuce under protective cover is an old practice. In earlier times the crops were grown in cold frames or hotbeds. Then greenhouses came into use (Fig. 4.9). At first, all were made of glass; later, most were covered with other materials such as various forms of plastic. Modern covered culture includes greenhouses, growing rooms, plastic tunnels and, for short periods after germination, flat plastic covering. Leaf and butterhead types are the most commonly produced items under cover, but some crisphead production is also practised.

Production of lettuce under cover is normally an activity for late autumn, winter and early spring in cold climates. Since the structure is completely enclosed, it offers the opportunity, not only to trap heat from sunlight, but also to modify temperature by heating, manipulate light duration and intensity, modify the atmosphere in which the plants are growing and control the water and nutrient supply. Light may be obtained from sunlight, sunlight

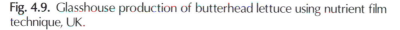

Fig. 4.9. Glasshouse production of butterhead lettuce using nutrient film technique, UK.

plus supplementary artificial light or artificial light alone.

Production of lettuce in greenhouses and similar structures grew as an industry in various parts of the world, particularly northern Europe, northern USA and Canada. Recently, however, when the cost of fuel became very high, growth of the industry slowed.

The simplest structure for protective cover is the plastic tunnel. Tunnels are used for temporary protection against low temperatures in early-spring plantings, when it is too cold to start plants without cover.

Greenhouses for lettuce production are usually very large, consisting of either ridge and furrow construction or single frame structures. The lettuce may be grown in soil or hydroponically. In the former case, plants are started in flats or in peat blocks in a section of the structure and then transplanted directly into the soil. The soil must be sterilized between crops to minimize disease. This is usually done with steam at 100°C. However, planting in freshly steamed soil can cause developmental problems, due primarily to excess uptake of manganese and changes in nitrogen availability (Sonneveld and Voogt, 1973). They found that sterilization with a steam–air mixture at 70°C killed all undesirable organisms and also released less manganese than at the higher temperature.

Most hydroponic production makes use of the nutrient film technique (NFT), with a variety of planting configurations. Plants may be grown in peat blocks on the floor in concrete troughs or in plastic troughs. Nutrient solution is introduced at one end of the structure and flows by gravity through the troughs in a thin film (3–6 mm). The solution is recirculated by pumping back to the upper end. Plants may also be grown on benches in growing trays with grooves or gutters at various spacings. The plants are held in perforated covers with their roots in the solution film. The solution is pumped to the head of each tray, flows down the gutters and is recirculated.

The most highly controlled structure is the growing room. Light is completely artificial and therefore completely controllable. Temperature levels can be regulated also, as there is no dependence upon trapped heat from sunlight. The growing procedure may be all-in all-out or continuous flow. In both cases the plants are grown hydroponically in nutrient solution. In a system where plants remain in place for the whole cycle, procedures are similar to those used in greenhouses.

The continuous growth system was developed in the 1960s. In this system, young seedlings are transplanted at close spacing at one end of a long tray. The plant is held upright with the roots in the tray, which contains a nutrient solution. By various means, the plants are moved towards the other end of the tray and the space between the rows of plants is increased as they grow larger. At the far end, the plants are harvested. In order for growth to occur at a predetermined rate for harvest, it is essential that the environment be as uniform as possible (Prince and Koontz, 1984). The growth rate can be calculated from the rate of weight increase to determine the space required for each plant at the time of harvest. Temperature and light intensity can then be adjusted to the time required for the plants to reach a harvestable size.

Several different types of lighting systems may be used, both for supplementary light in greenhouses and for total lighting in growing rooms. Traditionally, a combination of fluorescent tubing and incandescent lamps has been used to achieve near-daylight spectral distribution for photosynthesis and photomorphogenesis. In the 1970s, various types of high-intensity discharge lamps were developed. These included high-pressure sodium, metal halide and tungsten halogen lamps. Tibbitts *et al.* (1983) showed that the sodium and metal halide lamps alone or in combination and metal halide combined with tungsten halogen all gave acceptable growth characteristics. This occurred despite the fact that the sodium lamps produce little blue radiation (Table 4.1). Knight and Mitchell (1988) confirmed this with an experiment comparing the effects of incandescent plus fluorescent against metal halide on growth characteristics of Waldmann's Green leaf lettuce. Metal halide lamps are cheaper to use and this may be the main consideration in choosing a system.

Both in greenhouses and in growing rooms, the CO_2 concentration can be modified to enhance plant growth by increasing the rate of photosynthesis. Double or triple the normal atmospheric level will achieve this, as long as the increase in concentration is proportional to the light level. However, in winter, when the CO_2 increase may be obtained by venting flue gases from the heating

Table 4.1. Effect of four lamp treatments and two photosynthetic photon flux density (PPFD) treatments on five growth traits of lettuce (PPFD in μmol s⁻¹ m⁻²). (From Tibbitts *et al.*, 1983.)

Trait	PPFD	Metal halide plus tungsten halogen	Metal halide	Metal halide plus sodium	Sodium
Shoot dry weight (g)	700	1.36	1.86	2.00	1.98
	320	1.72	1.97	1.80	1.83
Plant leaf area (cm²)	700	427	546	621	693
	320	668	771	737	766
Leaf area ratio (cm² g⁻¹)	700	318	296	314	350
	320	393	391	411	418
Leaf number	700	12.5	13.3	13.0	12.8
	320	10.8	12.5	11.7	11.7
Hypocotyl length (mm)	700	0.80	0.50	1.10	1.30
	320	3.50	2.00	3.80	5.80
Chlorophyll (μg cm⁻¹)	700	6.10	8.43	7.11	3.81
	320	3.39	4.09	3.54	2.74

burners, oxides of nitrogen will also increase and may inhibit photosynthesis and therefore reduce yield (Caporn, 1989).

Greenhouse lettuce is a crop that is normally grown during the winter season when light conditions are poor, specifically in the northern regions of Europe and the USA. During this period in these areas, the radiation may be well under 300 cal cm⁻² day⁻¹, compared with a more desirable level of about 500. Glenn (1984) has shown that the radiation levels in sunny climates, such as desert locations, will produce heads of a given size in about half the time required in areas with poor light. He also found that light was used two to three times more efficiently in autumn than in spring. Daytime temperature is positively correlated with growth and interacts with radiation, which would explain the more efficient use of light in the warmer autumn season.

An understanding of the growth process in the winter period is important in order to optimize crop yield and may also be a useful breeding tool to enable selection of cultivars most likely to prosper and yield well under these conditions. Hunt *et al.* (1984) describe two procedures of integrated growth analysis as applied to butterhead lettuce grown in the greenhouse. Crop growth rate is analysed, in one procedure, in terms of relative growth rate and biomass and, in the other, in terms of the light interception and utilization efficencies of the plants. The curves generated were applied to compare winter growth characteristics of two butterhead cultivars and one crisphead (Hand *et al.*, 1985). Assessment of plant development under both methods was shown to be similar and they were both judged to be useful in analysis of dry-matter production.

Leaf lettuce is one of the crops under consideration for use in long, manned, space flights as a part of a regenerative life-support system. Energy-efficient light systems with optimum cost:benefit ratios are desirable and experiments have been performed using various systems alone and in combination. Development of efficient radiation sources with effective stable output has not yet been achieved (Mitchell *et al.*, 1991).

The growth of lettuce under greenhouse or growth-room conditions may also have some undesirable effects on crop quality. Of particular concern is nitrate content in leaves and stems. When consumed, nitrate may be converted to nitrite and then to compounds causing methaemoglobinaemia in infants, or to nitrosamines, which may be carcinogenic. The use of supplementary lighting or increasing photoperiod or photosynthetic photon flux all reduced nitrate concentration in greenhouse-grown lettuce (Gaudreau *et al.*, 1995). Santamaria and Elia (1997) showed that use of the ammonium form of nitrogen produced adequate yields of endive plants which were nitrate-free.

Reduction of the amount of nutrients provided may have a beneficial effect in reducing the eutrophication effect in discarded nutrient solution. Chen *et al.* (1997) found that lettuce growth and nutrient uptake were satisfactory when concentrations of N, P and K were reduced to 10% of commercial concentration in recirculating hydroponic systems and to 1% in non-recirculating systems.

Disorders of greenhouse-grown lettuce due to deficiency or toxicity of mineral elements is discussed, with colour pictures, in Roorda van Eysinga and Smilde (1981).

HARVEST AND POSTHARVEST METHODS

The objective of the sum of the various harvest and postharvest processes is to get fresh produce from the field to the consumer in good condition, preserving as much quality as possible. For lettuce, endive and non-forced chicory, the procedures are cutting, packing, cooling (and occasional short-term storage) and transporting the product to a market. These procedures must be done rapidly, with minimum damage to the product, so that it arrives with good quality. It then becomes the responsibility of the consumer who purchases the product to keep it in good condition until it is eaten. The key to accomplishing these goals is the cold chain, which means that throughout the course of these procedures the product is kept cool.

LETTUCE

Harvest Procedures

Nearly all lettuce is cut by hand. For most lettuce, harvest takes place at vegetative maturity. The small percentage that is used for mesclun is cut at a very early stage. The time of cutting is determined by the stage of maturity. This varies by lettuce type and by season, and is determined by inspection of the field. Crisphead lettuce takes longest to develop; maturity is signified by a firm head that gives only slightly to pressure. In the Salinas Valley, direct-seeded lettuce can be harvested in 60–65 days during the warmest part of the season. Summer lettuce in warmer climates may mature as rapidly as in 55–60 days. Lettuce grown through the coolest part of a winter season may require as long as 110–120 days to mature. Transplanted lettuce requires 3–4 weeks' less time to maturity.

Butterhead lettuce is mature when the head gives slightly to pressure and is usually 2–3 weeks earlier than crisphead types. Leaf lettuces are mature when the rosette is full and require about the same amount of time to mature as butterheads. Cos lettuces should have well-filled heads. Maturity time is earlier than with crisphead but not as early as in leaf and butter lettuces.

Lettuce is cut at the base with a hand-held knife, leaving the older green or senescing leaves on the ground. Then the head is trimmed of additional outer leaves. In the USA, head lettuce that is to be packed unwrapped is left with five to seven outer leaves, while that which is to be wrapped in the field is left with only one or two outer leaves. Other lettuces are trimmed of older, smaller and yellowed leaves. Heads are either placed back on the ground to be picked up and packed in a box, or they are placed on a vehicle with a conveyor belt.

Crispheads placed on the ground are for unwrapped, or naked, pack. The partially trimmed heads are picked up and placed in a cardboard carton (Fig. 5.1). Usually they are packed with 24 heads per carton in two layers. Large heads may be packed with 18 per carton and small ones with 30 per carton. In the USA, a decreasing proportion of the lettuce is cut, trimmed and packed in this manner. About 75% of crisphead lettuce is film-wrapped or bagged at the shipping point. The carton tops are closed and the cartons are placed on pallets on a trailer or an open-lorry, often 320 cartons per load, and are taken to a cooler facility (Fig. 5.2).

Fully trimmed heads are taken off the conveyor belt, film-wrapped or bagged in clear plastic, sealed and placed in cartons in various numbers, depending upon size. Most are packed at 24 per carton in the USA. In Europe, nearly all crisphead lettuce is wrapped. Heads are placed in smaller cartons in

Fig. 5.1. Crisphead lettuce harvesting and packing, Salinas Valley, California.

Fig. 5.2. Lettuce lorry being loaded with cartons of crisphead lettuce for transport to cooling facility.

one layer of 12 heads. In some cases, the heads may be placed in baskets first in the field and then transported to a shed for packing. Harvesting and handling in many countries is simpler and labour-intensive (Fig. 5.3).

Non-crisphead lettuces are packed in various-sized cartons, depending again upon size. In the USA, leaf and butterhead cartons contain 24 heads each and usually weigh about 11–12 kg. Cos cartons also contain 24 heads and weigh about 18 kg. In Europe these lettuces are packed at 12 per carton.

Film-Wrapping

Film-wrapping of lettuce has had a long, slowly developing history. The earliest attempt to package lettuce at the shipping point was by a single US handler in 1931, who wrapped the heads in cellophane (Chapogas and Stokes, 1964). Various other attempts were made in subsequent years, but it was not until the early 1960s that wrapping at the place of shipment became a commercial success (Fig. 5.4). Various wrapping methods, films, film shrinking procedures and wrapping facilities were tried. Several film characteristics are important. The film must be semipermeable to allow gas exchange (oxygen (O_2) and carbon dioxode (CO_2)) and allow evaporating moisture resulting from heat removal to escape, so as to prevent growth of rot-causing organisms. Excess permeability would allow too much moisture loss

Fig. 5.3. Partially harvested stem-lettuce field in Nile Delta, Egypt.

Fig. 5.4. Wrapped crisphead lettuce, Israel.

and wilting. The film must have a soft enough texture to have a pleasant feel to the consumer, unlike a number of early films, which were heat shrunk around the head and were stiff and crackled to the touch. It must take a brand label and be relatively inexpensive. Several types of materials and procedures have been developed to the present time. Flat films on rollers are cut, wrapped around the head and heat-sealed. In recent years, some shippers are placing heads in film bags, which are then closed with a clip.

There are several advantages to wrapping at the shipping point as opposed to transport of naked lettuce for wrapping at the receiving point. From 20 to 35% of the weight is removed with the outer leaves, which reduces transportation costs. The head and leaves are not forced into a tight place, thus preventing crushing and bruising. The head is also protected from damage by the film itself. The lettuce does not have to be trimmed of damaged leaves and wrapped at the retail level. A label printed on the wrap provides advertising for the lettuce company and the district in which it was produced.

Mechanical Harvesting

In the 1970s and 1980s, there was a great deal of interest in mechanical harvesting of lettuce. Experimental machines have been built and described in the USA and Holland. In the USA, machines developed in California and Arizona were selective harvesters for crisphead lettuce. The most advanced and tested machine was developed by the US Department of Agriculture Agricultural Research Service (USDA-ARS) in Salinas, California. This machine relied on a sensor to measure both the size and density of the head by the degree of attenuation of an X-ray beam passing through the head (Lenker and Adrian, 1971). Large dense heads were then cut with a hydraulically activated knife, grasped with rubber-fingered lifter belts and conveyed to the back of the machine for packing (Lenker *et al.*, 1973). This system did not become commercially useful primarily because the large volume of lettuce arriving at the back of the machine was difficult to place-pack in conventional cartons.

Non-selective harvesters have been developed and tested in Europe for both greenhouse butterhead lettuce and outdoor crisphead lettuce.

Packing Procedures

In many locations in the USA, the lettuce is inspected for quality characteristics during the early stages of harvest. The inspectors are hired by local authorities, but the costs of inspection are paid by the industry. Inspectors look for tipburn lesions or other blemishes, and are empowered to stop the harvest if the lettuce does not meet standards. In this way the industry safeguards the quality of the product shipped from that district.

The shape of the lettuce head, particularly of the crisphead type, is important for an efficient, high-quality packing process. Crisphead lettuce should be as nearly spherical as possible. Heads that are pointed at either the butt end or the top are more difficult to pack because they require extra space. Leaf types pack most easily if they form compact rosettes; this is also true of cos heads. Butterhead lettuces are smaller and softer and are relatively easily packed.

Crisphead lettuce may be harvested in one pass through a field if the crop has matured uniformly. However, maturity is variable and two or more cuts at intervals of several days may be required in order to harvest when heads are at optimum head firmness and density. Cos, leaf and butterhead lettuces are cut in one pass. Optimum shape and firmness are more easily obtained because greater uniformity is achieved when the growth and development period is shorter. Also, the criterion for head density is not as strict.

Maturity

The time to maturity of a lettuce crop is fairly predictable in regions where the climate is relatively uniform. In the western production areas of the USA, this condition usually exists. Based upon knowledge of soil types and microclimate of the various production locations, timing of plantings is made according to predictions of their maturity. These predictions are revised as the time of harvest approaches, but are remarkably accurate. Thus the grower can schedule plantings to ensure a steady supply of lettuce for harvesting.

The weather is, however, unpredictable in other locations. Wurr *et al.* (1988) studied the predictability of maturity in the UK, using 24 sets of data from 7 years of experiments with transplanted crisphead lettuce. They found that the best predictions could be made based upon effective day-degrees (EDD), using present and past weather records ($1/EDD = 1/DD + a/R$, where DD = day-degrees, a is a constant and R = total daily radiation (Scaife *et al.*, 1987)).

They believed, however, that predictions could not be based solely on weather data, but that developmental studies of the lettuce itself would be necessary in order to increase the accuracy of predictions. In a study on crisphead lettuce, Wurr *et al.* (1987b) showed that plants raised in the open and transplanted before the end of May were heavier and matured later than plants raised under glass at higher temperatures. In another study, Wurr *et al.* (1992) compared various statistical techniques to predict maturity and found that simple models based upon days, day-degrees and effective day-degrees were most accurate.

As much as 25% of the crisphead lettuce in a production area may be harvested in bulk for institutional or consumer packaging. In most cases, the lettuce field is contracted for this purpose, but, in many fields that are not completely harvested for carton packing, the remainder of heads in the field are sold for bulk harvest.

For bulk harvesting, the heads are cut and fully trimmed. The heads are then tossed into a bulk container for transport to a processing facility (Fig. 5.5).

Mesclun consists of very young leaves, with a maximum length of 10–12 cm. The leaves must therefore be cut carefully by hand, using a knife or shears. Each type of lettuce or other leafy vegetable is harvested and boxed separately. In some operations, the lettuces are seeded as a mixture and each of the other items is grown separately. The handler decides on the specific mix to be used. Since the leaves are very small and tender, it is essential to transport them to a cooling facility quickly to prevent wilting. The cutting is done above the crown, so that the plant will produce more leaves and can be cut two or three times.

Cooling and Transportation

Carton-packed lettuce is taken to a cooling plant, where it is unloaded by fork-lift off the lorry or trailer and placed on long narrow carts. These are wheeled into a vacuum tube for cooling. The doors are sealed and air is removed from

Fig. 5.5. Lorry trailer loaded with bulk cartons containing about 500 kg of crisphead lettuce for light processing.

the chamber with a vacuum pump. The lowered air pressure causes water to evaporate from the heads, sufficient to lower the temperature in the heads but not enough to cause wilting. The temperature is monitored and is lowered to 1°C. This process may take from 15–30 min, depending upon how much field heat must be removed from the lettuce. The cartons are then removed from the chamber. They may be immediately loaded into a transport vehicle or carried by fork-lift to a storage room, which is kept at 0°C and 98% relative humidity. The lettuce is usually stored for only a short period, until a lorry trailer is ready to be loaded. During the cooling cycle and in transport, it is essential that lettuce be kept away from certain ripening fruit or other sources of ethylene gas. Ethylene causes a condition called russet spotting in lettuce (see Chapter 3).

The first commercial plant for vacuum cooling was built in 1948 in Salinas. Many have been constructed since then, both permanent installations and portable types (Fig. 5.6). In the USA nearly all lettuce is vacuum cooled. In the UK, Spain and Israel all iceberg lettuce is cooled. In other countries, whether or not lettuce is cooled depends upon the distance from shipping point to market.

The theory and mechanics of vacuum cooling were discussed in a number of papers published in the early 1950s. A particularly good discussion of principles and types of units is provided by Isenberg and Hartman (1958).

Lorry trailers are refrigerated. The lettuce is ideally kept at about the same temperature as in the cold room, but sometimes, because of improper handling or faulty trailer design, the temperature may rise to 4 or 5°C. The lorries carry the product to terminal market cities. In the USA, a trip across the

Fig. 5.6. Small, portable, vacuum-cooling unit.

country from west to east may be 4800 km and may take 4 days. A trip from southern Spain to the UK may cover 2000 km and take about 2 days.

Bulk Harvesting and Consumer Packages

Bulk-harvested lettuce is carried to a processing plant. There it is cored, cooled, cut into various-sized pieces, washed, spin-dried and packaged. For food-service use, the lettuce is chopped or shredded and packed in bags holding several kilograms. It may be packed as lettuce only, or mixed with other salad vegetables, such as carrots and red cabbage. Food-service customers include restaurants, fast-food outlets, hospitals and schools. The lettuce is used in a coarsely chopped form for salads or in a shredded form for hamburgers or other types of sandwiches.

Lettuce in a consumer package is a rapidly growing marketing style. Lettuce is coarsely chopped and may be mixed with other leafy vegetables in different combinations and placed in clear plastic packages. The package has sufficient salad for one, two or more meals and may include crisphead, butterhead or cos lettuce alone or in combination, or one or more lettuces combined with radicchio, carrots and/or red cabbage. Some packages include small packages of croutons or salad dressing or both and others may contain a protein source, such as grated cheese.

Controlled-atmosphere (CA) storage has been found useful for both head lettuce and cut lettuce. Singh *et al.* (1972) found that maintenance of low temperature was of primary importance in preserving quality of head lettuce. However, CA storage produced added benefits. Lettuce was stored at 2°C, 2.5% CO_2 and 2.5% O_2 for 39 days; they found significant improvement in quality maintenance over storage under low temperature alone. The principal benefits of CA storage were reductions in decay, pink rib, butt discoloration and russet spotting (see Chapter 3).

There is concern that lettuce in cut form in various consumer and institutional packages may deteriorate because of the exposure of cut surfaces. Ballantyne *et al.*, (1988) studied the effect of modified atmospheres on shelf-life. They supplied the packages with an atmosphere of 5% O_2, 5% CO_2 and 90% nitrogen (N_2) and established an equilibrium of 1–3% O_2 and 5–6% CO_2. This treatment resulted in a 14-day shelf-life at 5°C, almost double that of the control packages. The packaging material was a 35 μm low-density polyethylene film. Use of a low-permeability film, with a low O_2 and high CO_2 atmosphere, resulted in anaerobic respiration, which produced undesirable odours.

Cut lettuce could be stored for 2–4 weeks if the initial quality was high, physical damage was minimized and the product was subjected to certain treatments (Bolin and Huxsoll, 1991). These include rinsing after cutting to reduce free cellular fluids, applying a slight vacuum and adding a small amount of carbon monoxide (CO). Browning could be retarded by dipping the lettuce in sodium dehydroacetate (Hicks and Hall, 1972).

ENDIVE

The handling of endive from harvest to postharvest is very similar to the handling of leaf lettuce. Endive, escarole and other salad items are usually grown in the same plantings as the non-crisphead lettuces and are usually harvested at the same time. Endive is cut and trimmed and usually packed in fibreboard cartons or wooden crates in the field. The cartons are transported to the vacuum-cooling facility to be cooled. After cooling, they are shipped in refrigerated lorries, which may contain mixed loads of several different non-crisphead lettuces and other leafy items that are available and ordered by a buyer.

CHICORY

Chicory other than the witloof type is handled in a similar manner to that for several of the lettuce types and endive. Specifically, the sugarloaf type, with an elongated head, is handled like cos lettuce. The spherical-headed radicchio type is treated like butterhead lettuce. The non-heading rosette forms are

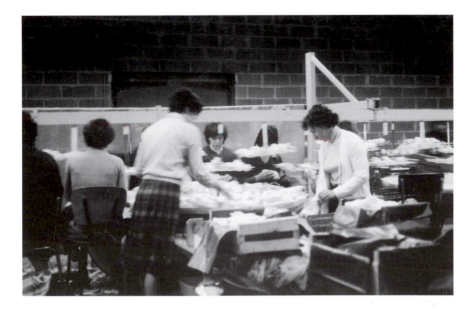

Fig. 5.7. Packing witloof chicons, France.

handled like leaf lettuce and endive. Containers may vary in size and shape. The overall handling procedure is essentially the same for all types.

Machine harvesting of radicchio is under study in Italy. Chicory roots for forcing are commonly machine-harvested in Europe. The machines are similar to potato diggers in principle.

Witloof chicons are harvested after 20–30 days of forced growth, as described in Chapter 4. The chicons are removed from the covering soil in the traditional method of forcing or lifted from the trays in the hydroponic method. As the roots are lifted, the chicons are hand-snapped from the roots, cleaned and trimmed as necessary. Chicons are then graded for size, compactness and shape and packaged according to grade standards (Fig. 5.7). In Europe, there are three market-grade classes, Extra, No. 1 and No. 2, used for both domestic and export markets. Those not meeting grade standards may be sold locally or not at all, depending upon price and quality.

The product is cooled by forced-air cooling, followed by cold storage, or placed directly in cold storage. Witloof should be kept at about 0°C and 95% relative humidity. The chicons should not be exposed to light, which induces undesirable greening.

6

SEED PRODUCTION AND MARKETING

The seed is the means of transition from one growing cycle to the next. Therefore, production and handling of seed is of major importance in the yield and quality of any crop. Seed is also a means of storage and conservation of a crop species.

LETTUCE

Day-length response and vernalization effects on lettuce flowering are discussed in Chapter 3.

Seed production is the last part of the second reproductive stage of the life cycle of lettuce. In earlier days, seed was harvested from plants that had already been cut for market lettuce. The head was cut high enough for axillary buds to remain and produce flowering stalks. Now commercial seed companies grow lettuce plants solely for the production of seed. Most lettuce seed is produced outdoors. A small proportion is produced under cover, which allows protection from wind, rain, insects and diseases.

There are a number of locations in the world where lettuce seed is produced commercially. The two principal areas are the west side of the San Joaquin Valley of California and the Griffith–Hay District in New South Wales, Australia. Seed is also produced in southern France and several other locations in the USA. California seed is sown in early May and harvested in September, while Australian seed is sown in late November and harvested in May. It is possible, therefore, to grow two seed crops in 1 year by growing in both places.

The sowing and growing of lettuce for seed is handled similarly to lettuce grown for market during the vegetative phase of growth. Non-crisphead cultivars are then simply allowed to produce a seed stalk; this stage is essentially unimpeded by the vegetative structure. Crisphead lettuce, however, forms a dense head. If allowed to grow into the reproductive phase, the stalk may not be able to elongate through the top of the head. In such a situation, the stalk may grow in a circular manner inside the head or it may break. One of several methods may be employed to prevent these occurrences. One is to slash the top of the head sufficiently to allow the stalk to come through easily. Alternatively, the top of the head may be sliced off. Finally, sharp downward pressure may be exerted on the head, which will break the leaf ribs and allow the head to be pulled off. Some lettuce fields are treated with gibberellic acid to encourage seed-stalk formation before heading takes place.

The seed fields are frequently inspected and rogued. At these times, plants are removed that are off-type, of a different cultivar or diseased, particularly if they appear to have lettuce mosaic virus (see Chapter 7).

Lettuce plants flower over a period of 50–70 days, but the flowering usually occurs in three peak periods (Jones, 1927; Soffer and Smith, 1974a). Seeds mature 12–17 days after flowering so that seed production occurs in peak periods as well. The characteristics of the seed harvested thus vary with the timing, method of harvest and number of harvests. Uniformity, yield, quality and subsequent performance may be dependent upon the conditions of growth as well as harvest techniques; these aspects have been the subject of much research.

Basic to the production of a high-yielding seed crop is the supply of moisture and nitrogen during the growing season. Hawthorn and Pollard (1956) found that high soil moisture was associated with high seed yield and slightly later harvest, but had no effect on seed viability. They found an increase of seed yield associated with increasing nitrogen up to about 90 kg ha^{-1} but not beyond that. Nitrogen application had no effect on viability or time of maturity. Izzeldin *et al.* (1980) found that sufficient moisture was needed during both the vegetative and reproductive growth stages to maximize seed yield. They also found, however, that moisture levels that were best for high seed yield produced smaller seed with lower vigour. An intermediate water deficit was found to lead to the best combination of yield and quality.

There appears to be a genetic component for seed yield. Hawthorn and Pollard (1951) found that strains of cv. Great Lakes lettuce selected for type also differed significantly in seed yield.

Temperature also has effects on seed yield, quality and subsequent performance of crisphead lettuce. Gray *et al.* (1988) grew cv. Saladin (Salinas) lettuce in a greenhouse at three day/night temperature regimes, 20/10°C, 25/15°C and 30/20°C. Day length was 16 h. The regimes were started when the plants began to flower. Seed yield increased and then decreased with successive increases in temperature. Temperature differences affected the number of mature florets per plant, seeds per floret and seed weight (Table 6.1). Increases in temperature reduced seed weight, which was associated with reduced germination. However, when imbibed at 30°C, only the seeds developed at the highest temperature regime germinated readily.

Table 6.1. Seed-yield traits of Saladin lettuce as affected by temperature regimes of seed-producing plants. (From Gray *et al.*, (1988.)

Crop	Temperature (°C)	Grams per plant	Thousand seeds per plant	Milligrams per seed
1	20/10	13.8	7.9	1.78
	25/15	30.4	30.1	1.07
	30/20	24.3	26.1	0.97
2	20/10	15.8	9.8	1.64
	25/15	22.8	23.6	1.04
	30/20	14.8	16.0	0.95

In a field trial with cv. Salinas lettuce, Steiner and Opoku-Boateng (1991) studied productivity and quality of the seed (number of seeds, seed mass and seedling-root length). They found a reduction in these traits at higher temperatures, although the germination percentage was increased.

Time of harvest affects seed quality. Globerson (1981) found that dry matter per seed increased for a period up to 12 days after anthesis. Seeds harvested 10–12 days after anthesis germinated at 100%, but germination was faster after 14 days.

Gibberellin has several effects when used on tight-heading cultivars. Application of gibberellin at the rosette stage increased the percentage of bolting and seed yield over no treatment or manual deheading (Harrington, 1960; Globerson and Ventura, 1973). Gray *et al.* (1986) found that gibberellin also increased uniformity of both bolting time and seed production, as compared with cutting the tops of mature heads.

There are two ways of harvesting lettuce seed. One is the 'shake method'. At intervals while the seed is maturing, the plant is bent over and the seed head shaken into a container so that the ripe seed falls off. This procedure has the advantage of minimizing loss of seed. It also allows the separation of seeds into lots based on time of maturity; the lots will vary in quality. The other method is to cut the plants down at a given time, allowing the incompletely mature seeds to develop further and then harvesting the seeds from the cut plants. Seeds may be lost this way and there is no opportunity to separate seeds based on time of maturity.

Seed Quality

Much research has been done on the nature and improvement of various aspects of seed quality, including germination percentage and rapidity, early growth or seed vigour, presence of disease, longevity in storage and harvest productivity.

O.E. Smith and colleagues conducted a number of studies on seed quality in the early 1970s. In the first paper of a series of five, Smith *et al.* (1973a) showed that seed weight was the best of several size indicators of seed vigour (Fig. 6.1). An assay called a slant test was developed to measure seed vigour. Vigour was differentiated from germination as a measure of emergence capability and was defined by radicle length 3 days after germination. Smith *et al.* (1973b) showed that seed vigour was positively related to rate of emergence, seedling size and subsequent growth, head size at harvest and percentage of marketable heads.

Soffer and Smith (1974a), in a study on flowering peaks, showed that seeds from the first two harvests were heavier than those from the third harvest. Over 90% of the seed yield came from the first 35 days of a 70-day flowering period. Seed size was not correlated with number of seeds per flower-head. Seed yield and quality were not affected by early harvest, withholding of water and nutrients in the last half of the flowering cycle or air temperature between 20°C and 34°C. In subsequent studies (Soffer and Smith, 1974b), they found that seed and embryo measurements influenced early vigour in the following order: weight > thickness > density > width > length. Influence of these traits on young plant fresh weight was significant, but negligible on marketable head weight (Fig. 6.2). They also (Soffer and Smith, 1974c) showed that increased nutrient levels in the soil increased seed weight but not vigour, while hydroponic culture increased seed weight but had a negative effect on germination and vigour.

Wurr and Fellows (1984) graded and primed (pregerminated) lettuce seed to study the effect on three traits. Priming increased radicle length but had no

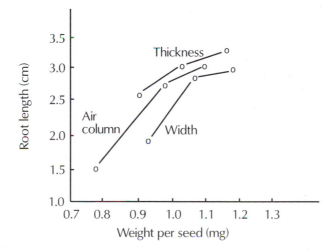

Fig. 6.1. Effect of seed weight of seed-lot fractions separated by three methods on seedling vigour (root length). For each separation, small (left), medium (centre), large (right). (From Smith *et al.*, 1973a.)

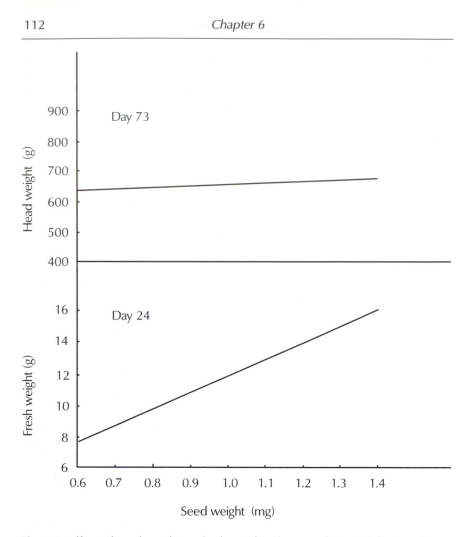

Fig. 6.2. Effect of seed weight on fresh weight of young plants (24 days) and on marketable head weight (73 days) of crisphead lettuce. (From Soffer and Smith, 1974b.)

effect on seedling size or maturity. Priming had no effect on germination at 20°C, but increased germination at 35°C. Priming in potassium orthiophosphate (K_3PO_4) was more effective than priming in water. Larger seed also increased radicle length and seedling size, but had no effect on maturity.

Storage-longevity conditions, treatments and consequences have been the subject of many studies. One of the earliest studies was by Kosar and Thompson (1957) on the effect of relative humidity (RH) on seed longevity. They found that under RH up to 67%, there was little or no effect on longevity of seed stored at 10°C. At 67% and 75%, seed stored for 3 or 4 years lost viability. Above 75%, viability was lost after only 1 year. Moisture content of the seed itself affected longevity (Barton, 1966). Seeds with moisture contents ranging from 5 to 50% were stored up to 18 years at ambient, 5°C, −2°C and −18°C. Seeds survived longer at the higher moisture content as the storage temperature was reduced. Conversely, survival under severe freezing temperature was increased as the seed water content decreased (Junttila and Stushnoff, 1977). Cultivar Grand Rapids lettuce seeds survived at −196°C (in liquid nitrogen) at moisture contents between 5 and 13%, but minimum survival temperature increased rapidly to −40°C as the water content increased from 13% to 16%. Roos and Stanwood (1981) found that high-moisture-content seeds would survived at least 33 days in liquid nitrogen, if the initial cooling was achieved rapidly. Freezing damage would occur if the temperature was decreased slowly. Vertucci and Roos (1990) studied the theoretical basis for establishing storage parameters and recommended that equilibration at 19–27% RH would provide optimum storage moisture.

Conditions of storage for lettuce vary depending upon the purpose of the storage. Storage in ambient conditions is acceptable for the short term only, as for seed-company stocks; the seed will lose viability in 1–3 years, depending upon the temperature and relative humidity ranges. For many years, the recommendation for medium-term to long-term storage was an environment where the temperature in degrees Fahrenheit plus the relative humidity did not exceed 100, and the temperature did not exceed 50°F (10°C) (Bass, 1973). The optimum storage for the long term is at −18°C and a low relative humidity. Storage at −196°C is effective for indefinite periods.

Villiers and Edgcumbe (1975) investigated the reason that fully hydrated seeds can lie for long periods in the soil and remain viable, but not germinate until an appropriate time. This is in apparent contradiction to the fact that viability decreases with increasing moisture in the low to medium or air-dry range. They found that lettuce seed of three cultivars showed increasing loss of viability with time when stored with moisture contents of 5.0, 7.0, 9.5 and 13.5%, with the loss of viability increasing faster at the higher moisture levels. However, fully imbibed seeds in darkness maintained 100% viability for 12 months. They concluded that ageing in the dry state allows accumulation of deleterious mutations and deterioration of lipids and protein membranes. Allowing the seeds to remain in contact with free water molecules prevents this deterioration and can reverse the effects of previous dry storage. Ibrahim *et al.* (1983) showed that, as moisture content increased under aerobic conditions, metabolic rates also increased. When sustained by respiration, repair of damage and replacement of damaged subcellular materials can take place under higher-moisture conditions.

An important measure of seed quality is the presence or absence of seed-borne disease agents. In particular, the presence of lettuce mosaic virus in lettuce plants at the flowering and seed-maturity stages can have significant effects: the production of seed is markedly reduced and the surviving seeds may carry the virus and serve as a source of primary infection when the seedlings emerge in a field planting (see Chapter 7). Other possible seed-borne pathogens are *Michrodochium panattoniana* (Berl) Magn., the cause of anthracnose; *Septoria lactucae* Pass.; *Pseudomonas cichorii* (Swingle) Stapp; and two other viruses, arabis mosaic and tobacco ring spot (George, 1985).

Seed Coating

Lettuce seeds are narrow and angular. Seeds are coated with various materials to give them a more spherical shape, which is more efficiently handled by planting machines (Zink, 1955). It then becomes possible to separate the seeds more effectively, which allows planting at a desired spacing between seeds. This means that considerably fewer seeds are needed to sow a unit area. In the thinning procedure, it permits removal of unneeded plants faster and more easily with less damage to roots and hypocotyls. The latter is particularly important in those areas, such as California, where short-handled hoes have been outlawed and the thinning crews have to use long-handled hoes and stand nearly erect. A disadvantage of coated seed has been that germination and emergence may be reduced, particularly in hot climates such as western deserts in the USA. However, improved coatings and other treatments have alleviated these effects. These include: (i) coatings that split easily; (ii) irrigation at night to take advantage of cooler temperatures; and (iii) seed priming.

Seed priming is a procedure to allow seeds to germinate under conditions that would normally be inhibitory, such as high temperature and presence of a seed coating. Seed priming consists of soaking seeds under conditions that permit them to imbibe water and initiate the germination process. The process must be halted before the radicle can penetrate through the pericarp. The seed can then be dried and stored for use when desired. Water, salt solutions or polyethylene glycol may be used for priming. High-quality seed is essential for successful priming (Perkins-Veazie and Cantliffe, 1984). Primed seed of cv. Empire germinated at 37°C, whether coated or not, after 5 months of storage at 5°C (Valdes and Bradford, 1987).

Seeds of cultivars sensitive to photodormancy can be exposed to light before coating to overcome a subsequent light-requiring situation. Also, coatings may include germination-enhancing chemicals, such as ethrel (K.J. Bradford, personal communication).

Some research has been done on creation of synthetic seeds (Sanada *et al.*, 1993). With lettuce, this can be done using adventitious shoots from tissue culture. These are encapsulated in alginate and can be of use principally in hydroponic culture.

Seed Marketing

Marketing includes a series of procedures. The seed must first be produced and then cleaned, stored, tested for germination, tested for virus presence, coated and packaged. It is then ready for sale.

For a new cultivar, seed production starts with the growing of breeder's seed, which is usually produced under greenhouse protection. Breeder's seed is used to produce foundation seed. This seed may be grown outdoors in progeny rows, which can be checked for uniformity and type. Foundation seed is then used to produce stock seed for sale to growers for market production. Stock seed is renewed about every 4 years. During that period, it is usually stored as bulk lots in bins under conditions as dry and cool as possible.

The seed is tested for germination percentage and should approach 100%. Lettuce seed must also be tested for seed-borne mosaic virus. Samples of 30,000 seeds are tested from each lot. If no virus is found, the seed lot is labelled as mosaic-tested or mosaic-indexed. If mosaic virus is found, the seed cannot be sold in areas where local ordinance demands seed that is essentially virus-free. Some lettuce-growing districts have their own testing facilities and will retest the seed received from the seed companies.

The seed is sized on an air-screen device and the portion with the lightest seeds is discarded. Nearly all seed is then coated with various materials. These usually have a diatomaceous earth or silicon-clay base and may also have additives designed to enhance germination and emergence. Coated, indexed seed can then be packaged in containers for sale, usually just before it goes on the market.

Seed of commercial cultivars may be sold directly to growers or sold first to seed dealers, who usually carry seed supplies from several companies. Cultivars for the packet, or home-garden, trade may be packaged by the company itself or sold to another company that specializes in that market. Seed in packets may vary considerably in germination ability, likelihood of virus presence and other aspects of quality, and is seldom coated.

ENDIVE

Many of the characteristics and production methods discussed for lettuce seed also apply to endive seed of annual cultivars. Seed should be sown in late summer or early autumn. To induce bolting, biennial cultivars need a cold treatment, either by overwintering or by a vernalization treatment (George, 1985). The main seed-borne pathogen of endive is *Alternaria cichorii* Nattr. (syn. *Alternaria dauci*), which causes black leaf spot.

Carette and Laurent (1989) compared plant architecture as influenced by plant density for the effect on seed yield, seed weight and germination. The simpler the architecture (greater density), the higher the yield and germination percentage, but the lower the seed weight.

CHICORY

Chicory is a biennial crop and must have a cold treatment to induce seed production. Chicory is largely cross-pollinated and therefore cultivars for seed production should be separated by a minimum of 5000 m (George, 1985). Pollination occurs by means of insect transfer (Rick, 1953). In southern Italy, seed is usually produced by individual growers, which has given rise to many landraces. The main seed-borne pathogens are *A. cichorii, Fusarium avenaceum* (Fr.) Sacc, *Rhizoctonia solani* Kuhn and chicory yellow mottle virus.

Little research has been done on chicory seed production. Fanizza and Damato (1995) analysed seed-yield components in two landraces of chicory grown in Italy. They found that the difference in seed yield between the two was due primarily to the number of fertile capitula.

7

DISEASES AND THEIR CONTROL

Diseases of lettuce, endive and chicory are caused by various organisms. They have the potential for two major economic effects. They may cause sufficient damage for the plant to be killed or rendered non-harvestable. This results in direct loss to the grower. The other consequence of disease may be to reduce the quality of the product. This may include blemishes and discoloration, flavour changes, reduction in size or changes in colour. Each of these problems has a tendency to depress the price, reduce shelf-life and make the product less desirable to consume.

Control of diseases is one of the great difficulties faced by growers and shippers. This is true for various reasons. One is that the symptoms of the disease may not be apparent early; when they appear, it may be too late to treat. Another is that the disease may develop suddenly, again making it impossible to treat. Third, there may not be a treatment available or, if available, it may be inadequate to achieve sufficient control to be economically beneficial. Fourth, for newly recognized diseases in particular, little may be known of the causal agent and its epidemiology, and it may be difficult to choose those aspects of the disease cycle that may be vulnerable to one or more control methods. Finally, the disease agent may have a means of adapting itself to the control method, such as developing resistance to a fungicide, so that it can overcome the control and become a recurring problem.

Fortunately, there are multiple ways of controlling many diseases and the grower has the option of choosing among them for the most effective means for a particular disease. These include the following.

- Treatment with a chemical substance.
- Treatment with a parasitic agent, such as a virus.
- Genetic resistance to the disease agent.
- Genetic resistance to symptom expression.
- Tolerance to presence of the agent.

- Non-inducing environment for abiotic disorders.
- Control of the vector for the agent.
- Selection of a growing environment unsuitable for the disease.

Diseases may be caused by a variety of agents: fungi, bacteria, viruses and virus-like agents, nematodes and abiotic causes. They may occur at various stages of the life of the plant: seedling stage, successive stages of growth and development or postharvest stage. They may affect different parts of the plant: roots, crown, stem, leaves, flowers, fruit and/or seeds.

DISEASES OF LETTUCE

The lettuce diseases described in this chapter are fully described, with coloured photographs, in the *Compendium of Lettuce Diseases* (Davis *et al.*, 1997).

Fungal Diseases

Downy mildew

Downy mildew is one of the most serious and most studied diseases of lettuce. It can occur in most locations where lettuce is grown and can be particularly troublesome in relatively cool, moist environments. It is a lesser problem in warm, dry climates. It can occur in any season, depending upon location. One to many pale yellowish lesions appear on the upper side of the outer leaves. The lesions are usually defined by veins, giving them an angular appearance (Fig. 7.1). Lesion size varies from about 0.50 cm × 0.25 cm to about 2 cm × 4 cm. Lesions may coalesce. They may appear on the cap leaves of heading lettuce. Several days after the lesions appear, sporulation takes place, usually on the underside of the leaf. The lesions later turn brown and may kill portions of leaves or whole leaves. Lesions on butterhead lettuces are likely to overlap the veins; subsequent damage is more likely to be serious than in the other types of lettuce.

The infection cycle begins when spores, which are wind-blown, alight on the surface of leaves. A spore will germinate if the temperature is 10–17°C, and if the relative humidity is close to 100%. On susceptible cultivars, the sequence of infection is: (i) germination of conidia on the leaf surface; (ii) formation of a germ tube and appressorium; (iii) direct penetration of an epidermal cell; (iv) invagination of the host plasmalemma during expansion of a primary vesicle in the cell; (v) development of a secondary vesicle; (vi) growth of hyphae intercellularly from the secondary vesicle; and (vii) formation of haustoria in cells adjacent to the hyphae. Penetration takes place within 4 h (Sargent *et al.*, 1973).

Particularly critical to germination is the duration of leaf wetness

Fig. 7.1. Downy mildew lesions on leaf of crisphead lettuce.

(Scherm and van Bruggen, 1994). They found that the average duration of leaf wetness in the morning of days on which infection occurred was 4.2 h, while the duration was only 1.9 h on days on which infection did not occur. After a short preincubation period, infection begins. The latent period, or time from infection to sporulation, lasts about 7–9 days. Sporulation takes place with the formation of conidiophores, which will only develop if there is a film of water on the leaf surface. Conidia are formed on the conidiophores in a few hours. Conidia become detached as the air dries and are dispersed by the wind to start the cycle again.

The agent causing downy mildew is a fungus, *Bremia lactucae* Regel. A number of forms, called physiological races, were identified. Each race was distinguished by its ability to parasitize a particular range of lettuce cultivars with specific resistance genes (Crute and Dickinson, 1976) (see Chapter 2). In recent years, isolates of downy mildew fungus collected in California were characterized for virulence phenotype, sexual compatibility and molecular markers (Schettini *et al.*, 1991). Based on this information, the isolates were classified into four pathotypes, emphasizing that they were dynamic, changing entities.

Bremia lactucae also has a sexual phase, which was poorly known until recently. Michelmore and Ingram (1980) demonstrated heterothallism in the fungus, and identified two compatibility types, B_1 and B_2. When the compatibility types were cultured together, abundant oospores were produced consistently. Only a few fungus isolates had both compatibility types. When heterothallic isolates of different virulence phenotypes were crossed, progeny were produced that exhibited novel virulence phenotypes (Michelmore and Ingram, 1981). This study also demonstrated the diploid nature of the fungus.

There are two principal methods of control of downy mildew. One is by the use of chemical treatments and the other is by use of resistant cultivars. Chemical control was only moderately successful against mildew, until 1978, when metalaxyl, a phenylamide fungicide, was introduced. It was highly effective until late 1983, when a field in England showed resistance of the fungus to the fungicide. Resistance appeared in the USA and elsewhere as well. It is due to a single gene (*Pp*), in which resistance is incompletely dominant (Crute and Harrison, 1988). Most *Dm* genes will not protect against metalaxyl-resistant strains of the fungus. However, a few do and can be used in conjunction with chemical control (Crute *et al.*, 1987). Combined treatment allows the sensitive portions of a population to be controlled chemically and the insensitive portions to be controlled by the appropriate lettuce gene. At the same time, the chemical will control any new sensitive phenotypes that arise.

Research is ongoing to allow more efficient control of downy mildew chemically, by means of a computer-based forecasting system. Its purpose will be to predict the likelihood of infection days immediately before their occurrence, at which time the grower can spray.

The other important means of control of downy mildew is with the use of resistant cultivars (see Chapter 2). Currently, cultivars carrying effective *Dm* genes are being used. However, this resistance has consistently broken down within a few years of the release of a new resistant cultivar. Resistance is dependent upon the action of a single gene. The organism can respond to the release of a cultivar with a new gene for resistance by mutating or recombining through sexual reproduction, creating a new virulent genotype, which can thus overcome the resistance.

Breeders are beginning to investigate the use of combined genes from exotic sources and the use of field resistance controlled by quantitative genes. These strategies are based on the theory that single-gene changes will not lead to immediate susceptibility. The field resistance does not confer immunity as do the *Dm* genes, but it provides a level of resistance that is usually adequate for effective economic control of the disease (Crute, 1984) (Fig. 7.2). Lesions usually develop slowly; they are usually small, confined to lower leaves and few in number. Tests conducted on leaf discs of 4–6-week-old plants showed that cv. Iceberg had non-specific resistance, manifested by a longer latent period, a lower disease intensity and reduced production of conidia (Gustafsson, 1992). The combination of quantitative and effective single-gene resistance would confer an adequate level of resistance and continued protection if the single-gene effect was lost.

Various forms of integrated control have been proposed in light of the

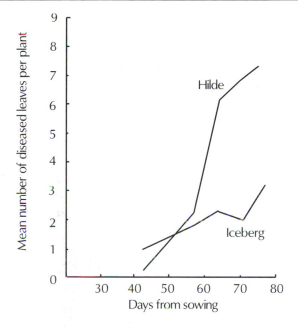

Fig. 7.2. Resistance of cv. Iceberg to downy mildew compared with susceptibility of cv. Hilde. (From Crute, 1984.)

incomplete or partial control obtained with present methods used singly. Crute (1984) proposed three possible combinations: single-gene resistance plus field resistance, single-gene resistance plus fungicides and field resistance plus fungicides (Fig. 7.3).

Sclerotinia drop

Sclerotinia drop has been reported in many places where lettuce is grown and is probably a worldwide problem. An attack on seedlings causes a rapid collapse and drying. On older plants, the first symptom after invasion of the fungus is a wilting and flattening of the lower leaves, usually starting on one side of the plant. Subsequent leaves and finally the whole plant wilt and flatten, with increasing yellowness and necrosis, until the entire plant is dead (Fig. 7.4). There may be a brown soft rot at the crown. A white cottony mycelium appears on the undersides of leaves and at the crown. The mycelium bears the hard, black, irregularly shaped, resting bodies, of various sizes, called sclerotia.

The disease can be caused by two related species, *Sclerotinia minor* Jagger or *Sclerotinia sclerotiorum* (Lib.) DeBary. They are distinguishable in several ways. *Sclerotinia minor* forms small sclerotia, 1–2 mm in diameter, while *S. sclerotiorum* produces sclerotia that may be over 2 cm in diameter. Also, *S. minor* produces ascospores infrequently and they play little or no part in causing infection (Grogan *et al.*, 1955). *Sclerotinia sclerotiorum* produces

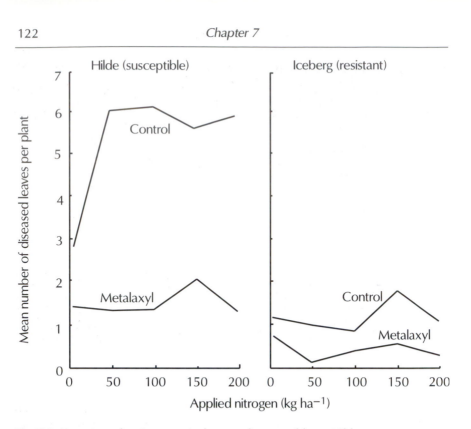

Fig. 7.3. Reaction of resistant cv. Iceberg and susceptible cv. Hilde to treatment with metalaxyl. (From Crute, 1984.)

apothecia on its sclerotia, which in turn produce ascospores. These can be wind-blown and cause infection at some distance. *Sclerotinia minor* can infect only nearby plants, when a sclerotium germinates eruptively close to the stem or root, so that the emerging hyphae touch the nearby plant. *Sclerotinia minor* is more common in the Salinas Valley in California, while *S. sclerotiorum* is found most often in the San Joaquin Valley, the Imperial Valley and Arizona. The Santa Maria Valley appears to have approximately equal amounts of both forms. This variation in distribution pattern also occurs in other parts of the USA and in Europe and may be dependent upon soil type and temperature. Cultivation or conditions that reduce soil moisture will prevent apothecia formation by *S. sclerotiorum*, but not the vegetative growth of *S. minor* (Beach, 1921; Abawi and Grogan, 1979). Also, *S. sclerotiorum* is a greater problem in wet seasons than in dry ones. *Sclerotinia minor* spreads more slowly and is more localized, but recurs more consistently, and will cause disease in periods of fluctuating dryness and wetness. In the literature, there is some confusion as to which species was being observed. There is also some disagreement as to whether they should be considered one species or two (Purdy, 1979).

   Research on survival and activity of *S. minor* in the soil has been

Fig. 7.4. Severe wilting of crisphead lettuce plants with *Sclerotinia* drop.

conducted on both the east and west coasts of the USA, with both similar and differing results. Imolehin and Grogan (1980) found that inoculum densities in the field varied from place to place in the Salinas Valley. Sclerotia survived better in drier than in wet soil and better at shallow compared with greater depth. They also survived better in soils with a history of drop than in soils with no such history. Disease intensity was proportional to inoculum density. Imolehin *et al.* (1980) found that the optimum temperature for sclerotial germination and mycelial growth was 18°C, with a range of 6–30°C. Abawi *et al.* (1985) compared soils with and without a history of drop for effect of water potential. Survival decreased as water potential increased, regardless of the soil type. They concluded that the difference in a soil's ability to support the disease is based upon level of biological activity. Adams (1987) found lower survival in drier soil, but found that increasing temperature decreased survival. However, the range of temperatures (35–50°C) was considerably higher than the maximum studied by Imolehin *et al.* Adams found that survival was greater at greater depth, which may be explained by differences in experimental technique or in soils.

Drop may be controlled in several ways: use of fungicides, biological control, cultural control and disease resistance. Timing and placement of materials are important for the success of a fungicidal treatment (Steadman, 1979). The material should be applied before infection occurs. Particularly with *S. minor*, the material should be placed at the plant–soil line. Treatments with fungicides vary in effectiveness with the chemical used. The post-thinning period is the best time for control, and it is essential to have a continuous layer on the soil surface (Marcum *et al.*, 1977).

Biological control of sclerotia of both types has been demonstrated with the mycoparasite *Sporidesmium sclerotivorum* Uecker, Ayers and Adams (Ayers and Adams, 1979). Adams and Ayers (1982) evaluated the parasite under field conditions and found that it reduced the numbers of sclerotia of *S. minor* by 75–95%, depending upon the level of inoculation (Table 7.1). Control of *S. sclerotiorum* was reported by Budge *et al.* (1995). Application of the fungal antagonist *Coniothyrium minitans* Campbell reduced the number and viability of sclerotia.

Mulching with black polyethylene film significantly reduced the incidence of *S. minor* on lettuce (Hawthorne, 1975). Two other mulches had no effect. Removal of infected lettuce plants from a field for 3 consecutive years reduced inoculum level and incidence of drop (Patterson and Grogan, 1985). Deep ploughing can reduce the density of *S. minor* sclerotia and incidence of drop in the next crop, but is not effective on subsequent crops (Subbarao *et al.*, 1996). The ploughing redistributes the inoculum in a more uniform manner, resulting in the infection of a higher proportion of plants. Addition of barnyard manures suppresses the activity of *S. sclerotiorum*, possibly due to the increased activity of antagonistic organisms (Asirifi *et al.*, 1994).

Breeding for resistance to *S. minor* on lettuce is in its beginning stages. However, some work has been done on the existence of genetic resistance. Newton and Sequeira (1972) identified 21 Plant Introduction (PI) accessions of various types with a higher level of resistance than Great Lakes 659. Some of these lines were used in crosses with susceptible material; the F_2 populations suggested a genetic basis for resistance. Elia and Piglionica (1964) found significant differences in susceptibility among named cultivars of lettuce.

Table 7.1. Disease incidence and control of *Sclerotinia* drop of lettuce with *Sporidesmium sclerotivorum*. (From Adams and Ayers, 1982.)

	Spring 1980		Autumn 1980	
Conidia (no. g⁻¹ soil)	Percentage plants with symptoms	Percentage disease control	Percentage plants with symptoms	Percentage disease control
0	64	0	24	0
1	60	6	20	15
10	49	23	15	38
100	36	43	10	59
1000	22	65	6	75

Powdery mildew

The first authenticated report of powdery mildew on lettuce was in 1941 in the Salinas Valley (Pryor, 1941), on a breeding line derived from a cross between a susceptible wild lettuce and a resistant crisphead lettuce. It reappeared in 1951, on cultivated lettuce, and was considered a different strain from that found by Pryor and originally restricted to wild lettuce (Snyder *et al.*, 1952). It has been reported occasionally since then and has also been reported in several countries in Europe. It usually occurs as a white mycelium on both sides of older leaves. The mycelium occurs in non-circumscribed areas, unlike downy mildew. Leaf areas damaged by powdery mildew become yellowed and then necrotic.

The disease is caused by *Erysiphe cichoracearum* DC ex Merat, an ascomycete. It may occur in at least two strains, one infecting wild lettuce and one infecting cultivated lettuce. Optimum condition for germination of conidia is 18°C (range 6–30°C) at less than 100% relative humidity (Schnathorst, 1960). After infection of the leaf tissue, the mycelium emerges on both surfaces to sporulate in about 4 days (Deslandes, 1954). Perithecia of the sexual stage may appear on the leaves. The disease may occur as early as April in the Salinas Valley but is most common in the autumn.

Although powdery and downy mildews each have different optimum conditions for their occurrence, these overlap sufficiently for both diseases to be able to occur at the same time in the same area, even on the same plant (Schnathorst, 1962). In the Salinas Valley, usually only powdery mildew occurs in the warmer, south part of the valley, and only downy mildew in the north part, while both can occur in the large central part of the valley.

Control of powdery mildew can be achieved with application of sulphur dust; the temperature must be sufficiently high to volatilize the sulphur (Grogan *et al.*, 1955). No resistance breeding has been carried out, although sources of resistance are known. In a test of 46 cultivars of various types, only butterhead and cos cultivars showed resistance; none of the crispheads or leaf types tested were resistant (Schnathorst and Bardin, 1958). Resistance to the strain that infects wild lettuce (Pryor, 1941) was shown to be controlled by a single dominant allele from the cultivated parent (Whitaker and Pryor, 1941). Resistance also occurs on some accessions of *Lactuca saligna* and other species of *Lactuca* (Lebeda, 1994).

Anthracnose

Anthracnose is also known as shothole disease and as ringspot. It is caused by a fungus, *Michrodochium panattoniana* (Berl.) Sutton *et al.* (formerly *Marssonina panattoniana*). It has been found in various parts of the world and in various lettuce areas of the USA. The disease occurs in cool, wet seasons, such as in the early spring, or in the summer in cool locations.

Plants are infected by splashing drops of rain or sprinkler irrigation water. Conidia can infect through the stomates or through the epidermis (Couch and

Grogan, 1955). Symptoms appear as small circular chlorotic lesions, which then become tan or brown. These are usually on lower leaves near the soil surface. Later, the centres of the lesions become necrotic and drop out. Infected leaves may dry up and be blown away. The organism overwinters in previous crop-lettuce debris in the soil as microsclerotia; these can survive up to about 3 years (Patterson and Grogan, 1991).

The organism can survive on leaves of lettuce during shipment and, when inoculated on mature lettuce, it will develop symptoms after 10–17 days at 7–18°C (Moline and Pollack, 1976).

Several fungicide materials reduced incidence of the disease but did not eradicate it (Parman *et al.*, 1991). Genetic resistance exists. Ochoa *et al.* (1987) screened 449 cultivars and PI lines for resistance to five races of *M. panattoniana*. They found that Salad Bowl, a leaf lettuce, was resistant to three races; only one item, an accession of *L. saligna*, was resistant to all five.

Bottom rot

Bottom rot is a fungal disease of lettuce that occurs primarily in a warm humid environment. In the USA, this includes the eastern and Midwest lettuce-growing districts. It is less common in cooler, drier areas. The disease is caused by the fungus *Rhizoctonia solani* Kuhn. Infection begins with the germination of microsclerotia in the soil under the leaves touching the ground. These leaves become infected, and this is indicated by the appearance of discrete necrotic lesions. Additional larger lesions form and the fungus also progresses upward through the leaves. These may drop off. The fungus eventually penetrates the heads, which become small, black and mummified. The entire course of the disease may take as few as 10 days (Townsend, 1934). Infection doesn't usually start until the plants are approaching maturity, when the lower leaves cover a substantial area of ground. However, it may begin as early as 4 weeks after planting.

The problem may be exacerbated by the secondary invasion of rotting organisms, such as *Erwinia carotovora* and *Pseudomonas marginalis.* These cause severe rotting and sliming of the lettuce. They enter the plant through the tissue damaged by the *R. solani* (Pieczarka and Lorbeer, 1975).

There are at present only two useful methods of control. One is by the use of fungicides; the other by ridging (Pieczarka and Lorbeer, 1974). Fungicide spray should be applied to cover the underside of the bottom leaves to reduce the disease incidence. Planting on flat ground has traditionally been the common practice on the muck soils of the east and Midwest of the USA and in Europe. Ridging raises the lower part of the plant above the soil surface, thus suppressing the action of the fungus.

Stemphylium leaf spot

Stemphylium leaf spot has been identified in several countries. However, the only lettuce-production area in which it is considered a serious problem is

Israel, primarily because it may reduce the amount of crop available for export (Netzer *et al.*, 1985). The disease was identified and described in the USA by Pahdi and Snyder (1954). It occurs primarily during wet, cool weather.

The disease is caused by *Stemphylium botryosum* Wallr., the imperfect form of *Pleospora tarda* E. Simmons. The form that infects lettuce was named *S. botryosum* f. *lactucum* by Padhi and Snyder (1954). Conidia germinate on the leaf surface. The germ tube grows until it can penetrate the leaf through a stomate, after which a mycelium is produced. The typical symptom is a necrotic spot, which may be rounded or angular. The spots may enlarge and coalesce.

All cultivars tested by Padhi and Snyder were susceptible. Netzer *et al.* (1985) could find no sources of resistance in *Lactuca sativa*, but found resistance in a population of *L. saligna* collected in Israel. They concluded that resistance was conferred by the dominant allele of one gene and the recessive allele of another. The resistant wild lettuce was crossed with cultivars in order to transfer this resistance. No other forms of control have been found to be useful.

Botrytis

This disease is also called grey mould. It can infect lettuce in greenhouses and occasionally in the field, but it is primarily a postharvest problem (see Chapter 3).

Verticillium wilt

Verticillium wilt, unknown on lettuce until a recent outbreak in a restricted growing area, is highly damaging and has the potential to be a serious disease of lettuce (Subbarao *et al.*, 1997). The disease is well known on a number of crop species, including cotton, tomatoes, strawberries and cauliflower. The disease is caused by *Verticillium dahliae* Kleb, a slow-growing fungus. The first symptom on lettuce is a wilting of the lower leaves, as early as the late rosette stage. Cutting the root discloses a greenish black discoloration of the vascular system. At head formation of crisphead lettuce, the plants turn yellow, and the outer leaves become dry and appressed around the head. Microsclerotia appear on the veins of the outer leaves.

The ultimate extent of the disease is not yet known. No control methods have been developed. Genetic resistance is a potential control.

Other fungal diseases

Several other fungi have been shown to be pathogenic to lettuce. Several *Pythium* spp. cause rotting, or damping off, of germinating seedlings of lettuce, endive and chicory. It has been reported by Stanghellini and Kronland (1986) that *Pythium dissotocum* Dreschs is a subclinical feeder on lettuce rootlets and may be responsible for yield losses.

Other organisms have been reported once or a few times and several are summarized in Table 7.2.

Table 7.2. Other fungal diseases.

Disease	Organism	Reference
Downy mildew	*Plasmopara lactucae-radicis* Stanghellini and Gilbertson	Stanghellini *et al.*, 1990
Rust	*Puccinia dioicae* Magnus	Chang *et al.*, 1991
Leaf spot	*Centrospora acerina* Hartig	Neergard and Newhall, 1951
Wilt	*Fusarium oxysporum* fs. *lactucum* nov. spec.	Hubbard and Gerik, 1993
Leaf spot	*Cercospora longissima* Sacc	Savary, 1983
Black root	*Chalara elegans* Nag Raj and Kendrick	O'Brien and Davis, 1994

Bacterial Diseases

Corky root

In the past, corky root was attributed to various environmental conditions and organisms. Among these were phytotoxic lettuce debris from a previous crop, overfertilization with certain types of fertilizers and invasion by *Pythium ultimum* Trow. The causal organism was recently identified as a Gram-negative bacterium (van Bruggen *et al.*, 1990c). The organism is *Rhizomonas suberifaciens* van Bruggen, Jochimsen and Brown.

The disease is first manifested by the appearance of small yellow-brown lesions on the roots (Fig. 7.5). These become larger and darker, especially on the tap root, which then becomes brownish green, rough and cracked. The smaller roots may be destroyed and the tap root may be pinched off near the crown. The plant may develop new feeder roots near the crown. In a severe infection, virtually all the roots are destroyed. The upper part of the plant responds to severe root loss with yellowing, wilting, reduced head size and even death. The interior of the root may show brown discoloration. This symptom can also indicate fertilizer damage (see Chapter 4).

Corky root has been reported in California, New York, Florida and Wisconsin in the USA. It occurs on muck soils in the three eastern states and on mineral soils in California. It has also been reported in Canada and Italy. In California, it is a problem in the coastal districts, Salinas and Santa Maria Valleys, but not elsewhere in the state or in Arizona. In those locations where the disease occurs, it can have severe effects, reducing head size and yield. O'Brien and van Bruggen (1992) found yield reductions of up to 92% for the susceptible cultivar Salinas, but no significant reduction for a resistant breeding line (Table 7.3). Datnoff and Nagata (1992) had similar results in Florida.

On mineral soils, the disease is more severe on heavy-textured soils, particularly if they are compacted and remain moist.

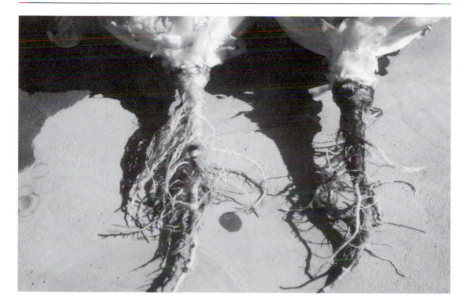

Fig. 7.5. Reaction to corky root of resistant plant (left) and susceptible plant (right).

Corky root can be controlled by various field-management practices, by chemical treatment and with the use of resistant cultivars (see Chapter 2). Rye as a winter cover crop reduced the severity of corky root but did not consistently improve yields (van Bruggen *et al.*, 1990b). The effects may be related to reduction in soil moisture by the roots of the rye crop. Alvarez *et al.* (1992) proposed that crop rotation would be useful for minimizing losses from corky root in Florida, based upon comparison of yield on land previously cropped to sugar cane and land continuously cropped, for 3 or 6 years, to lettuce.

Table 7.3. Reaction of susceptible and resistant lettuce to corky root incidence in microplots at Davis, California. (From O'Brien and van Bruggen, 1992.)

Season	Lettuce line	Treatment	Percentage diseased	Percentage marketable yield	Head weight (g)
Autumn	Salinas	+	99.5	8	94
	(suscep.)	−	0.4	90	431
	440−8	+	37.9	97	434
	(resis.)	−	0.8	96	437
Spring	Salinas	+	75.6	65	324
	(suscep.)	−	2.1	94	558
	440−8	+	17.2	98	567
	(resis.)	−	0.0	96	578

Varnish spot

The name varnish spot was used by Salinas Valley growers to describe a disease that was first seen there in the 1970s. A disease similar to varnish spot has been identified in Japan (Hikichi *et al.*, 1996).

Varnish spot is caused by a rod-shaped bacterium, *Pseudomonas cichorii* (Swingle) Stapp. Varnish spot is characterized by shiny, firm, dark brown spots, which are found on leaf blades and petioles below the outer two or three head leaves (Fig. 7.6). It occurs on crisphead lettuce in fields that have been sprinkler-irrigated up to harvest, but not in fields that were furrow-irrigated after thinning. The disease appears on nearly mature plants. Since the symptoms are not visible on outside leaves, the field cannot be selectively harvested. In this respect, it is a field problem similar to tipburn (see Chapter 3).

The strains of *P. cichorii* that caused varnish spot on mature lettuce do not cause disease on seedlings, nor do they cause soft rotting symptoms. The pathogen is soil-borne and probably infects plants by splashing caused by irrigation sprinklers.

Three bacterial diseases described as head rots were studied by Burkholder (1954) in New York. *Pseudomonas cichorii* was one of the bacteria implicated, but the soft-rot symptoms described were different from those of varnish spot

Fig. 7.6. Symptoms of varnish spot on head leaf of crisphead lettuce.

(Grogan *et al.*, 1977). The differences between varnish spot and head rot may be due to a difference in strains, an environmental difference between New York and California, or confusion in the New York study between the effects of *P. cichorii* and *P. marginalis*.

Since the disease in California does not occur on lettuce that is furrow-irrigated during the maturation stage, this practice would appear to be a useful control method. No other control methods have been described for varnish spot.

Bacterial leaf spot

Bacterial leaf spot is caused by a Gram-negative rod-shaped bacterium, *Xanthomonas campestris* pv. *vitians* (Brown) Dye. It was reported in New York on lettuce by Burkholder (1954) and later in Florida and California. It has also been reported in Italy, Brazil, Australia, Japan and South Africa. The infection is manifested as black angular lesions on the leaves of lettuce. These spots may coalesce to form large areas of damage. The pathogen is transmitted by splashing of rain or irrigation water and may also be seed-transmitted.

No methods of control have been developed. Differences have been observed in susceptibility of cultivars in the field, but no definitive study has been reported.

Other bacterial diseases

Various other bacteria have been ascribed as causes of disease in lettuce. The two most commonly reported are *P. marginalis* (Brown) Stevens and *E. carotovora* (Jones) Holland. Both cause soft rots.

Virus Diseases

Lettuce mosaic

Lettuce mosaic was first identified as a virus disease by Jagger (1921) in Florida. Subsequently, the disease has been found in lettuce-production areas all around the world. Losses as high as 100% have been recorded. In certain locations, severe losses would occur year after year until control measures were developed in the late 1950s and early 1960s.

Lettuce mosaic virus is a member of the potyvirus group. It is a flexuous filamentous rod (Couch and Gold, 1954). The virus is insect-transmitted in the field and can be sap-transmitted in the laboratory. It is also transmitted from one generation to the next through infected seed. It is likely that the rapid movement of the disease around the world was due to shipments of seed lots containing seed-borne virus. Infected seeds produce primary inoculum in a small proportion of emergent plants.

The first symptom on infected plants in the seedling stage or young rosette stage is vein-clearing, which appears systemically 10–14 days after inoculation. There is also a slight rolling back of the margins of the leaf, showing the vein-clearing symptom. Subsequent leaves show a mottled effect, caused by loss of chlorophyll in patches on the leaf, giving a mosaic of light and dark tissue (Fig. 7.7). Necrotic symptoms may appear on certain cultivars. Growth is restricted so that the plant becomes more and more stunted as it approaches maturity. At the same time, the mottling effect tends to be lost and the plant appears light or dull green, with the leaf margins turned downward, giving a wilted appearance. If the plant is allowed to bolt, the cauline leaves and the involucres of the flower-head develop necrotic lesions. Seed formation is drastically reduced.

A small percentage of seeds on an infected plant will contain the virus in the embryo. This percentage is usually 1–3%, but may range up to about 15%

Fig. 7.7. Typical mottling symptoms of lettuce mosaic on lettuce leaf.

(Couch, 1955). The later the infection, the less likelihood of seed-borne virus. An infected seedling from a virus-containing seed may show distorted cotyledons as it germinates, and mottling symptoms will appear on the first or second true leaf.

As a field problem, lettuce mosaic can be extremely destructive. Whole fields may be lost, and, in areas where lettuce occupies a significant portion of the land, multiple fields may be lost. Infection can take place in two ways. If a seed lot contains a proportion of infected seeds, each of these seeds that emerges then becomes a primary source of inoculum for further infection. Secondary transmission occurs when an insect vector feeds upon an infected plant. The most common vector is *Myzus persicae* Sulz., known as the green peach, or peach-potato, aphid. The virus is transmitted in a non-persistent manner. The aphid in feeding picks up virus particles on its stylets. Then, after withdrawal the aphid feeds upon a healthy plant, transferring the virus particles into the cells of the healthy plant. Large aphid flights are most common in the spring and autumn in many locations. At these times, large numbers of aphids move from plant to plant as they feed, and the virus is easily spread through the field. In places where lettuce fields are planted in sequence over an entire season, aphids may move from one field to a younger one, thus infecting successively planted fields and causing great losses. This was most notable in the Salinas Valley (Grogan *et al.*, 1952) and in south-east England (Broadbent, 1951) before control methods became available.

Other aphids, such as *Pemphigus bursarius* L. (McLean, 1962) and *Macrosiphum barri* Essig (Dickson and Laird, 1959), can transmit lettuce mosaic virus, but none are as efficient and effective as *M. persicae*.

There are a number of other hosts of lettuce mosaic virus (Costa and Duffus, 1958) (Table 7.4). Hosts include weed and crop species and several recently identified ornamentals, including *Gazania* spp. (Zerbini *et al.*, 1997) and *Osteospermum fruticosa* (L.) Norl. (Opgenorth *et al.*, 1991). Overwintering plants of some of the weed species can act as sources of infection through aphid feeding and transmission. Any host species growing on ditch banks, in home gardens and yards or on the edges of lettuce fields can act as a source of inoculum.

The stage of growth at which a field is infected with lettuce mosaic determines the amount of economic damage caused at harvest time (Zink and Kimble, 1960). Plants of Great Lakes 118 were inoculated at three stages of growth. The earlier the inoculation, the greater the reduction in plant size and yield. If a whole field is inoculated at an early growth stage, the lettuce is likely to be unmarketable at harvest. Reduction in economic value decreases with later infection.

Severe isolates of lettuce mosaic virus have appeared occasionally. Recently, such isolates have been reported in Greece, Yemen and Spain. Pink *et al.* (1992a) compared these and two additional strains, from England and the USA, against six crisphead cultivars. They were able to divide the strains into three pathotypes, two of which had not been identified previously. Pathotype I

Table 7.4. A list of species known to be hosts of lettuce mosaic virus. (From Costa and Duffus, 1958, and additional sources.)

Weeds

Anagallis arvensis	Chenopodium urbicum	Medicago polymorpha
Capsella bursa-pastoris	Cicer yamashitae	Nicotiana clevelandii
Carduus broteroi	Cirsium vulgare	Picris echioides
Carduus pycnocephalus	Erodium cicutarium	Rumex britannica
Chenopodium album	Lactuca livida	Senecio vulgaris
Chenopodium amaranticolor	Lactuca saligna	Silybum marianum
Chenopodium ambrosioides	Lactuca serriola	Sonchus asper
Chenopodium capitatum	Lactuca virosa	Stellaria media
Chenopodium murale	Lamium amplexicaule	Urospermum picroides
Chenopodium quinoa	Malva parviflora	

Agronomic plants

Carthamus tinctorus	Cichorium intybus	Spinacia oleracea
Cicer arietinum	Lactuca sativa	Tetragonia expansa
Cichorium endivia	Pisum sativum	

Ornamental plants

Aster spp.	Gomphrena globosa	Senecio cruentus
Callistephus chinensis	Lathyrus odoratus	Tagetes erecta
Chrysanthemum maximum	Osteospermum	Zinnia elegans
Gazania spp.	fruticosum	

included the strains from Yemen and Greece, pathotype II included the strains from the USA and England and pathotype III included the strain from Spain. Cultivar Salinas was susceptible to all pathotypes. Cultivar Ithaca was resistant only to pathotype I. Cultivar Salinas 88 was resistant to I and II and susceptible to III, but with a delay in symptom development. Cultivars Calona, Malika and Vanguard 75 were also resistant to I and II, but did not react to III with the delay in the susceptibility reaction. The authors suggest the use of these cultivars as a differential set. Subsequently, another strain of the virus was found on endive (Dinant and Lot, 1992), to which Malika was susceptible, while Vanguard 75 was resistant.

Control of lettuce mosaic was difficult to achieve for many years. Attempts to control the spread of the virus by treating with insecticides were not successful. The virus particles are transferred on the aphid's mouthparts. As soon as the stylets penetrate the plant, transfer of the virus occurs, even though the insect is quickly killed by the insecticide. The seed-borne effect also hindered control. Infected seeds were distributed over the entire field; the plants that emerged from those seeds were infected. The infected plants acted as foci for infection centres from which the rest of the field could be inoculated by aphid movement.

The seed-borne aspect of the disease plays a major role in one of the two methods of control that have been developed successfully. Newhall (1923) suggested that seed free of the virus would provide a means of control by eliminating the primary infection centres. Grogan *et al.* (1952) showed that the use of mosaic-free seed for plantings in commercial fields substantially decreased the incidence of disease and increased the yield. They proposed the use of mosaic-free seed as a means of control. Zink *et al.* (1956) showed that, as the percentage of seed-borne virus decreased, the number of plants that became infected in the field planted with the seed decreased. However, at seed-borne percentages greater than 0.1%, the degree of control was not acceptable (Table 7.5).

As a result of this research, seed indexing became mandatory for planting lettuce seed in most growing areas of the USA and common practice in many areas in the world. The most common standard is no infected seeds in 30,000 seeds. This became the standard for seed sold in the USA. In addition to the use of virus-free seed, in Monterey County and in some other districts, growers are expected to plough old fields, clean up host weeds on roadsides and field edges and avoid scheduling of new plantings immediately adjacent to old fields. These procedures have been highly successful and have reduced the incidence of infection to low, manageable proportions (Grogan, 1980).

In early years, seed lots were tested for presence of virus by observing seedling symptoms. Later, a test plant, *Chenopodium quinoa* Willd., which reacts to infection by mosaic virus with development of local and systemic symptoms, was used. Now, a procedure called enzyme-linked immunosorbent assay (ELISA) has been adapted for detection (Falk and Purcifull, 1983). The indexing procedure has worked well and has eliminated lettuce mosaic as a major problem in most districts in the USA.

Resistance was identified in the cultivar Gallega (Von der Pahlen and Crnko, 1965) and later in three primitive lettuce introductions from Egypt (Ryder, 1968). Gallega was used in breeding programmes in Europe as a source of resistance, while one of the lines from Egypt was used in programmes in the USA. Since the European growers have just recently adopted

Table 7.5. Effect of decreasing amounts of seed-borne lettuce mosaic virus (LMV) on proportion of plants with lettuce mosaic at first harvest. (From Zink *et al.*, 1956.)

Trial	\multicolumn{8}{c	}{Percentage seed-borne LMV}
	1.6	0.8	0.4	0.2	0.1	0.05	0.025	0.0
1	27.5	8.9	7.6	4.2	3.6	1.5	1.5	1.4
2	19.4	9.2	8.4	7.1	4.8	2.7	2.6	1.2
3	22.5	15.7	13.2	12.2	11.3	7.8	8.4	4.4
4	14.2	8.0	7.4	3.6	2.6	1.4	2.3	1.3
5	47.5	35.6	34.3	18.2	11.5	8.8	3.9	5.2
6	82.9	75.9	77.7	47.6	23.1	17.3	11.6	8.3

seed indexing for control, resistant cultivars appeared fairly early after the identification of resistance. In the USA, the indexing procedure became well established and resistant cultivars appeared at a slower pace. The first resistant cultivar was Vanguard 75, released in 1975 (Ryder, 1979b).

The resistance is conferred by a single recessive allele, which drastically reduces the rate of multiplication of the virus in the plant. Therefore, although the infection becomes systemic, symptom expression is greatly restricted and there is little effect on the growth or appearance of the plant. The inheritance of resistance was studied by Bannerot *et al.* (1969), who designated resistance in Gallega as *gg*, and by Ryder (1970), who assigned *momo* for resistance in the primitive lettuce. It was believed that they were the same until Dinant and Lot (1992) concluded that there were two resistance alleles at the same locus (Table 7.6). Pink *et al.* (1992b) revised the pathotype designation to accommodate this difference. Pathotype III, to which all alleles are susceptible, became pathotype IV. The endive strain, which affects Malika (*gg*) and Vanguard 75 (*momo*) differentially, became pathotype III.

Each control method used alone is useful in reducing the incidence of mosaic in lettuce plantings. There is value as well in using both methods. Ryder (1973) has shown that seed transmission occurs in resistant lettuce, although at about a 90% lower rate than in susceptible lettuce. This attribute will be useful in indexing programmes, since it would reduce the number of seed lots of both resistant and susceptible cultivars likely to contain seed-borne mosaic and thus permit more lots to be accepted for planting.

Big vein

Big vein was first reported in California in 1934 (Jagger and Chandler, 1934). It has subsequently been reported elsewhere in the USA, England, western Europe, Australia and other areas where lettuce is grown. In the field it is characterized by two primary symptoms. One is a clearing of the laminar area around the veins, giving an appearance of enlarged veins – hence the name of

Table 7.6. Reaction of differential cultivars to four pathotypes of lettuce mosaic virus (LMV). Vanguard 75 may also have the Ithaca resistant gene. (From Pink *et al.*, 1992b.)

Cultivar	LMV pathotypes			
	I	II	III	IV
Saladin (Salinas)	S	S	S	S
Ithaca	R	S	S	S
Malika	R	R	S	S
Salinas 88	R	R	R	S
Vanguard 75	R	R	R	S

S, susceptible; R, resistant.

the disease (Fig. 7.8). The other is a tendency for the leaves to assume a stiff, upright appearance and ruffled margins. Delay of heading and reduction in head size may also be associated with big-vein infection.

Big vein is a soil-borne disease. It is transmitted to lettuce plants by a root-feeding fungus, *Olpidium brassicae* (Wor.) Dang. (Grogan *et al.*, 1958). Transmission is by motile zoospores; the fungus survives as resting spores in the soil. The disease can be transmitted from lettuce to lettuce by grafting, but not through mechanical means (Campbell *et al.*, 1961; Tomlinson and Garrett, 1964).

Fig. 7.8. Vein-clearing symptom of big vein on lettuce leaves.

For many years, the agent of the disease was unknown. The discovery that it was graft-transmissible indicated that it was virus-like. However, it was called big-vein agent rather than big-vein virus, because it could not be further characterized. Later, the presence of virus-like particles was shown by use of the electron microscope and through aetiological means (Kuwata *et al.*, 1983; Vetten *et al.*, 1987). In addition, Mirkov and Dodds (1985) identified double-stranded ribonucleic acid (RNA) in the roots of infected lettuce, which suggested that big vein is caused by an RNA virus. Confirming evidence that big vein is virus-induced will come with the completion of Koch's postulates.

Big-vein symptom expression is temperature-related (Westerlund *et al.*, 1978a). Translocation from root to top was most rapid at 18–22°C. However, symptom expression was most severe when the air temperature was 14°C, whether the roots were at 14°C or 24°C. Symptom expression was virtually eliminated if the air temperature was 24°C, whether root temperature was 14°C or 24°C. Subsequent work by Walsh (1994) showed that symptom expression occurred faster and with greater severity at 18°C than at 14°C.

Symptom expression is also affected by soil type (Westerlund *et al.*, 1978b). Soils can be classified as big-vein-prone or big-vein-suppressive. Big-vein-prone soils have a heavy texture and hold water well, thus permitting movement of the infective zoospores. Suppressive soils are light and they dry quickly, hindering zoospore movement.

The effects of temperature and soil type act to encourage symptom expression in the field during winter and early spring growing periods. Expression is suppressed during warmer growing periods, especially on light-textured soils. The fungus remains in the soil for long periods. During warm periods, the agent may be transmitted into the root and subsequently trans-located to the leaves, but symptoms are absent.

Big vein has been identified in lettuce plants grown for the first time in various locations. Since it is only soil-transmitted, this suggested that other species, including weeds, could harbour both the fungus and the agent. Campbell (1965) found big-vein-bearing isolates of *O. brassicae* in related *Lactuca, Cichorium* and *Sonchus* spp. Some showed symptoms of big vein, others did not.

Big vein can be controlled chemically, culturally and genetically. The most effective chemical control is by soil fumigation with methyl bromide (Campbell *et al.*, 1980). Fumigation controlled *Olpidium* and reduced big-vein incidence in two succeeding crops. Also, the crops were earlier and more uniform. However, methyl bromide fumigation is expensive and is being phased out of use. In recirculating nutrient solutions, several fungicides and surfactants will control big vein (Tomlinson and Faithfull, 1979).

In cool, moist periods, lettuce should be planted on light, sandier soils that dry and warm easily, rather than on heavy soils with good moisture-holding ability. This becomes less important as the ambient temperatures rise.

A low level of resistance was identified in the cultivar Merit (Thompson and Ryder, 1961). This cultivar was used in breeding cultivars with greater resistance: Thompson, Sea Green and Pacific. Further screening has disclosed additional cultivars with resistance (Bos and Huijberts, 1990; Ryder and Robinson, 1995). Apparently certain lines of the wild species *Lactuca virosa* are immune (Bos and Huijberts, 1990).

The internal nature of resistance is unknown, but is manifested as a reduced proportion of inoculated plants showing symptoms at a specific stage of maturity. In field plantings, this is measured at market maturity. In greenhouse tests, the proportion of plants expressing symptoms after an arbitrary number of days from planting or transplanting is the criterion.

Beet western yellows

Lettuce is affected by several viruses and virus-like entities that cause yellowing diseases. Among these diseases are beet western yellows, lettuce infectious yellows, lettuce chlorosis, aster yellows, necrotic yellows, beet pseudo-yellows and beet yellow stunt. They have in common a degree of resemblance of symptom type, but vary in vector, host range, geography and economic importance.

Beet western yellows was first named radish yellows (Duffus, 1960, 1961). The virus causes an interveinal yellowing of the lower leaves and sometimes the outer head leaves of lettuce. It has a wide host range, which includes 146 species in 23 genera. The virus is transmitted primarily by the green peach aphid (*M. persicae* Sulz.). It is not seed-transmitted in lettuce.

In lettuce, the disease causes stunting as well as interveinal yellowing on lower leaves. The yellowing may progress over most of the surface of the leaf. The virus may also cause symptoms on younger leaves. Older infected leaves become brittle and thickened.

The disease is widespread; it has been reported in most places where lettuce is grown and is probably common around the world (Duffus and Russell, 1972). The economic importance of beet western yellows on lettuce varies with lettuce type and other factors. On crisphead lettuce there is not usually enough damage to reduce the harvest. Occasionally, the yellowing extends far enough into the head leaves for losses to be sustained. In a greenhouse study, beet western yellows and lettuce mosaic viruses alone or together caused substantial reductions in several traits, including seed production (Ryder and Duffus, 1966). The virus has been reported as damaging in several European countries, in both reduced yield and marketability. It is considered the most damaging virus to lettuce in the UK (Walkey and Pink, 1990). These losses occur primarily on butterhead lettuce, which exhibits a more severe symptom expression than crisphead cultivars.

No effective control of beet western yellows on lettuce has been developed. Differences in susceptibility among cultivars were noted by Watts (1975). Walkey and Pink (1990) developed a screening procedure and identified several cultivars as having some resistance, including butterhead, Batavian and iceberg types. In general, they found that crisp cultivars were less susceptible than butterheads. As a group, the Batavian types showed the greatest level of resistance. Two wild species, *Lactuca perennis* and *Lactuca muralis*, were virtually immune to the virus.

Two studies described genetic resistances to beet western yellows. Pink *et al.* (1991) crossed two previously identified resistant cultivars, Crystal Heart (Batavian) and Bursc 17 (butterhead), with a susceptible butterhead, Dandie. It was suggested by F_2 data that resistance was inherited as a single recessive gene. No F_3 data were taken. Maisonneuve *et al.* (1991) crossed a resistant and a susceptible accession of *L. virosa* and found from F_2 and first back-cross (BC_1) data that resistance was due to a single dominant gene. In both programmes, crosses to crisphead lettuce cultivars were made to transfer the resistance genes to commercially useful types.

Lettuce infectious yellows and lettuce chlorosis

Lettuce infectious yellows is a virus disease that became economically important in the desert districts of California and Arizona in the 1980s. However, its importance since 1990 has become uncertain. It was first noted in the 1981/82 autumn–winter season, accompanied by large populations of the sweet potato whitefly (*Bemisia tabaci* Gennadius). It was shown that the disease was caused by a virus and vectored by the whitefly (Duffus *et al.*, 1986). The virus induced interveinal yellowing symptoms, as well as stunting, rolling, vein-clearing and brittleness of affected leaves (Fig. 7.9). Leaf necrosis may also develop. The disease affects a variety of other crops, ornamentals and weed species. Klaasen *et al.* (1995) showed that the agent is a closterovirus and reported the complete nucleotide sequence.

Beginning in 1990, the sweet potato whitefly was nearly wholly replaced in 3 years by a new form, either a new strain of the insect or a new species called silverleaf whitefly (*Bemisia argentifolii*) (Perring *et al.*, 1993) (see Chapter 8). This whitefly was a poor transmitter of lettuce infectious yellows, which also virtually disappeared. However, a new virus that causes similar symptoms to infectious yellows appeared in the desert plantings. It is also a closterovirus, transmitted by both forms of the whitefly. The yellowing disease has been named lettuce chlorosis (Duffus *et al.*, 1996). The potential for damage by lettuce chlorosis is not yet known.

In a study by McCreight *et al.* (1986), differences in tolerance to lettuce infectious yellows were found among lettuce cultivars, and, in a second study (McCreight, 1987), resistance was identified in *L. saligna* accessions. In the first study, the cultivar Climax, although not resistant, showed less severe symptom expression than the other cultivars in the test. No cultivars have yet been developed from these sources.

Other yellowing diseases

Beet yellow stunt is caused by a flexuous rod-type virus, transmitted primarily by the sowthistle aphid (*Nazonovia lactucae* L.) (Duffus, 1972). Symptom

Fig. 7.9. Lettuce infectious yellows. Severely infected plants of cv. Empire on left, tolerant plants on right.

expression in lettuce starts with chlorosis and folding back of rosette leaves, followed later by stunting, severe yellowing and necrosis. Brown necrosis of the phloem tissue is diagnostic.

Beet pseudo-yellows is caused by a virus and transmitted by the greenhouse whitefly (*Trialeurodes vaporariorum* Westwood) (Duffus, 1965). It is a greenhouse disease, although it has been identified on weeds growing in the wild. The symptoms are interveinal yellowing of the older and intermediate leaves.

Lettuce necrotic yellows has been found only in Australia and New Zealand. It is a virus disease, vectored by the sowthistle aphid (*Hyperomyzus lactucae* L.) (Stubbs and Grogan, 1963). The symptoms of lettuce necrotic yellows resemble those of tomato spotted wilt (Fry *et al.*, 1973). Plants infected at an early growth stage become dull green, with bronzing and necrosis along the veins. Heads are stunted and unmarketable. Plants infected at later stages may develop necrosis of the veins and of the innermost leaves.

Sowthistle yellow vein is caused by a virus transmitted by the sowthistle aphid (Duffus, 1963). The symptoms include vein-clearing and banding. In lettuce the virus shortens the tips of the leaves, giving them a truncated appearance (Duffus *et al.*, 1970).

Tomato spotted wilt
Tomato spotted wilt virus causes disease in a number of species, including lettuce. The virus has a very broad host range, including vegetables like tomato and celery and ornamentals such as nasturtium and dahlia. It is found on lettuce in diverse locations around the world, including Hawaii, California, South Australia, South Africa and Chile.

Lettuce plants infected when young turn yellow, flatten and die. Older plants develop marginal wilting, yellowing and necrosis of the leaves. The damage tends to be one-sided (Fig. 7.10). This appears to cause a lateral curvature of the midribs of the less affected leaves. The youngest leaves have numerous brown-black spots, or the entire interior portion of the heart may turn black. The virus causing spotted wilt is transmitted by species of thrips. It

Fig. 7.10. Tomato spotted wilt. Leaves on near side more severely affected than those on far side. (Courtesy of R. Bardin.)

may also be transmitted mechanically in the laboratory. The western flower thrips (*Frankliniella occidentalis* Pergande) is the most common vector on lettuce in Hawaii, which is one of the areas most seriously affected by the disease. Several other species of the Thripidae are also vectors (Cho *et al.*, 1987).

Spotted wilt in lettuce has been very difficult to control. Elimination of weed and ornamental hosts, disc-harrowing of old fields and planting in less infested areas have been suggested for control (Harris, 1939). In Hawaii, some growers have been forced to end lettuce production in certain areas where spotted wilt was severe (Cho *et al.*, 1989). In the absence of effective resistance in useful cultivars, several integrated procedures were developed by a team assembled to construct a multidisciplinary approach. The procedures were proposed for the precrop, crop and postcrop phases. The first-phase procedures are: (i) crop rotation with non-susceptible species; (ii) planting susceptible crops adjacent to resistant ones; and (iii) control of alternate hosts of the vector. During the crop phase, procedures are: (i) use of virus-free seedlings; (ii) use of insecticides; and (iii) reduction of cultivation to reduce thrips movement. For the postcrop phase, they are: (i) fallow areas with high disease incidence for 3–4 weeks to allow the thrips to leave; and (ii) fumigate the soil. Crop placement and cultivation reduction have produced limited good results.

Research is in progress for the development of resistant cultivars (O'Malley and Hartmann, 1989). Two butterhead cultivars, Tinto and Ancora, were identified as moderately resistant compared with a susceptible check. They were crossed with each other and with the susceptible cultivar Green Mignonette. Results indicated that the two resistant cultivars had the same resistance genes and resistance was partially dominant. Lines from another source of resistance, a cross between *L. sativa* and *L. saligna*, are also being used in a breeding programme (Wang *et al.*, 1992).

Cucumber mosaic

Cucumber mosaic is a virus disease found on lettuce in the UK, Europe, North America and New Zealand. It is particularly damaging in autumn plantings of the lettuce crop in New York (Provvidenti *et al.*, 1980). Because of the development of lettuce mosaic virus-indexed seed, cucumber mosaic has replaced lettuce mosaic as the most important virus problem in New York. Symptom expression caused by the two viruses in lettuce are similar: vein-clearing and leaf recurving, followed by mottling and sometimes necrosis. Plants become stunted and distorted. Symptom expression of cucumber mosaic virus is milder than that of lettuce mosaic virus and the virus is not seed-borne in lettuce. It is transmitted primarily by the green peach aphid (*M. persicae* Sulz.). It has a very wide host range, including a number of weed species that appear to be natural sources of infection in lettuce (Bruckart and Lorbeer, 1976), of which *Barbarea vulgaris* R. Br., *Ceracium arvense* L. and *Rorippa islandica* (Oeder) Borbas appear to be the most important overwintering sources in New York. Most control measures are ineffective; resistance seems to be the most likely means of control. A source of resistance has been found in a line of *L. saligna* (Provvidenti *et al.*, 1980), but only to one of two known virus strains.

Other virus diseases

Broad bean wilt is often found associated with cucumber mosaic and/or lettuce mosaic. It is found less commonly than cucumber mosaic in New York, but is now more frequently identified than lettuce mosaic (Provvidenti *et al.*, 1984). The virus causes diseases in several species and is transmitted by several aphid species, especially the green peach aphid. Symptom expression is very similar to that of cucumber and lettuce mosaics. A number of lines of *L. virosa* and cultivars of lettuce have been found to have tolerance, which is expressed as a mild symptom or a virtually symptomless reaction.

Among virus diseases, *Bidens* mottle is second in incidence to lettuce mosaic in Florida (Zitter and Guzman, 1974). Symptom expression is similar to that of lettuce mosaic, but more severe. It is not seed-borne, but is transmitted by aphids, especially by the green peach aphid. The virus was first found on the weed species *Bidens pilosa* L. and *Lepidium virginicum* L. (Christie *et al.*, 1968), which are likely to be the sources for infection of lettuce (Zitter and Guzman, 1974). Reduction of the disease can be achieved by elimination of the host-weed populations. Also, recessive resistance to one of two identified strains has been found in the cos cultivar Valmaine.

Turnip mosaic virus was first identified in lettuce in 1966 (Zink and Duffus, 1969). The virus is transmitted principally by the green peach aphid, but is not seed-borne. Turnip mosaic is characterized by numerous circular to irregular lesions on the leaves. Severe stunting and even lethality may also occur. In reproductive stages, the virus causes necrosis on the cauline leaves and involucres. The disease is considered a serious problem in France. A lettuce cultivar survey showed association with downy mildew: in some cultivars, resistance or susceptibility to both diseases was found. It was later shown that *tu* was closely linked to *Dm-5/8* (Zink and Duffus, 1969, 1970; Robbins *et al.*, 1994).

Several other viruses occur sporadically in lettuce. Lettuce speckles disease is caused by two viruses acting together. One is the beet western yellows virus. The other, which has been named lettuce speckles mottle virus, fails to cause symptoms when alone (Falk *et al.*, 1979). The disease is characterized by small angular yellow spots on outer leaves. It has only been found in California. Dandelion yellow mosaic attacks lettuce as well as dandelion (Kassanis, 1947). It has been found in England, Holland and Czechoslovakia. *Arabis* mosaic virus has been identified in England on lettuce (Walkey, 1967). It causes severe stunting, chlorosis and necrosis and appears to be seed-borne. Lettuce calico has been ascribed to tobacco ringspot virus by Grogan and Schnathorst (1955) and to alfalfa mosaic virus by Stone and

Nelson (1966). It is not clear that the same disease was discussed by both pairs of authors. Lettuce leafroll mosaic, a disease caused by a caulimosaic virus-like agent, has been reported in Taiwan (Chen and Chen, 1994). It infects only *Lactuca* species and is characterized by mosaic symptoms, downward leafrolling and stunting.

A second disease vectored by *O. brassicae* has been known in Holland, France, the UK and Belgium since the early 1980s and has recently been identified in California (Campbell and Lot, 1996). It is called lettuce ring necrosis, kringnecrose or maladie des taches orangées. It affects butterhead and Batavia-type lettuces most severely and may be virus-induced.

Sonchus yellow net virus was found on lettuce in Florida (Falk *et al.*, 1986). Chicory yellow mottle virus has been reported on lettuce in Italy (Vovlas and Quacquarelli, 1975), and tobacco rattle virus on lettuce in California (Mayhew and Matsumoto, 1978).

Aster Yellows — a Phytoplasmic Disease

Aster yellows is a disease that affects approximately 350 plant species in 54 genera. It was first reported on lettuce as the Rio Grande disease (Carpenter, 1916). Kunkel (1926) reported occasional extensive losses to the disease in New York and identified the aster or six-spotted leafhopper *Macrosteles quadrilineatus* Forbes (formerly *Macrosteles fascifrons* Stal.) as the vector. In the eastern states, the disease did not infect celery or zinnia. However, in California, Severin (1929) was able to infect both species and proposed that two strains existed. Later, these were named the eastern and western strains. The western strain is transmitted by several leafhoppers. At that time the organism causing the disease was considered to be a virus. Later work showed that the cause was a phytoplasma-like organism (Errampalli *et al.*, 1991).

Aster yellows in lettuce first appears as blanching and yellowing of the young leaves (Fig. 7.11). Pink or tan latex deposits are found around the heart leaves, which may be stub-like in appearance. The plant is stunted, fails to form a head and is unharvestable. The flower-head has a bushy appearance, and immature flowers are distorted. The disease can only be transmitted by the leafhopper or by grafting. The disease is rarely serious on lettuce in California, but can cause severe losses in eastern states and countries in Europe.

There are no known sources of resistance to aster yellows in lettuce. Control of the vector can be attained with frequently repeated spray applications. A study by Lee and Robinson (1958) compared eight cultivars of lettuce for tolerance to aster yellows. Of the eight cultivars, only one, Trianon Cos, showed some tolerance.

Fig. 7.11. Aster yellows. Healthy plant on left, infected plant on right, showing yellowing and rolling back of leaves.

Nematodes

Nematodes can damage lettuce grown in warmer climates and soils. Lettuce is normally grown during cooler periods. Therefore, nematode problems are relatively minor in commercial lettuce crops. However, five forms of nematodes have been documented as pathogens on lettuce. These consist of two *Meloidogyne* spp., *M. hapla* Chitwood and *M. incognita* (Cofoid and White) Chitwood, *Pratylenchus penetrans* (Cobb) Sher and Allen, *Longidorus africanus* Merny and *Rotylenchus robustus* (de Man) Filipjev.

The root-knot nematodes (*Meloidogyne* spp.) affect a large number of species of crop plants all over the world. The northern root-knot nematode (*M. hapla*) is widely distributed in New York (Wong *et al.*, 1970). Soil fumigants were shown to be effective in reducing the incidence of nematodes on lettuce in organic soils. The southern root-knot nematode (*M. incognita*) has been reported as a problem on lettuce in Costa Rica (González and López, 1980). They found that disease incidence was related to soil type and inoculum density.

The meadow lesion nematode (*P. penetrans*) has been identified on lettuce in Italy, England and elsewhere. *Longidorus africanus* was identified on lettuce as a pathogen in the Imperial Valley of California. *Rotylenchus robustus* is pathogenic to several cultivars of lettuce in California.

DISEASES OF ENDIVE

Weber and Foster (1928) reported the occurrence of several diseases on endive and escarole as well as on lettuce. These included: *Sclerotinia* drop, downy mildew, damping off, anthracnose, grey mould, bottom rot and other fungus-incited diseases; several bacterial diseases; lettuce mosaic; and tipburn. In general the symptom expression of these diseases is the same as in lettuce. Falk and Guzman (1981) reported that a disease of endive and lettuce in Florida, called spring yellows, appears to be related to beet western yellows. Big vein is the principal lettuce disease that has not been reported on endive types.

Lettuce is planted more widely and has greater economic worth than the endive types, and therefore most disease problems are more economically damaging on lettuce. One exception is *Bidens* mottle, which occurs on lettuce in Florida, but affects endive and escarole more severely (Zitter and Guzman, 1974). The effects of *Bidens* mottle and lettuce mosaic are very similar in lettuce, and they are not easily distinguished from each other visually. However, *Bidens* mottle symptoms are more severe than mosaic symptoms on endive than on lettuce, and they can be distinguished visually. In Florida, *Bidens* mottle occurs more frequently on endive, while lettuce mosaic is more frequent on lettuce. Zitter and Guzman (1977) tested several endive cultivars and PI lines for reaction to *Bidens* mottle virus. All were susceptible.

Turnip mosaic is a virus disease that affects endive as well as lettuce. Several endive cultivars were tested for reaction to turnip mosaic virus (Provvidenti *et al.*, 1979). All were susceptible. Another new virus, escarole mosaic virus, was described on escarole in southern Italy (Crescenzi *et al.*, 1996). It causes mosaic, yellowing and necrosis. It appears to be transmitted by seed and probably by pollen, but not by any of five insect vectors tested. It can infect lettuce, several other species of *Asteraceae* and species in other families.

DISEASES OF CHICORY

Chicory is also affected by many of the same diseases and disorders as lettuce. There are, however, a number of diseases that have been reported specifically as chicory problems. Schoofs and De Langhe (1988) described four fungal diseases that have been reported only for chicory. Violet root rot, incited by *Helicobasidium brebissonii* (Tul.) Pat., produces a reddish purple mycelium on the root and causes a rot of the outer layers of the root. *Phytophthora erythroseptica* Pethyb. causes a black or brown rot at the distal end of the root. *Phoma exigua* Desmaz. produces black or brown spots on the root and restricts water uptake so that the head is reduced in size. *Puccinia cichorii* (DC) Bell, causes rust. Spore pustules are produced, which kill affected leaves quickly.

Verticillium albo-atrum Reinke and Berth. (*V. dahliae* Kleb) causes yellowing, wilting and marginal browning of the leaves and discolours the interior of the root. These symptoms are similar to those incited by *V. dahliae*, which has been newly reported on lettuce (Subbarao *et al.*, 1997).

During the witloof forcing period, the most important disease organisms are: *P. erythroseptica*, *P. marginalis*, *P. exigua*, *E. carotovora* and several fungal rotting organism.

Two viruses have been identified on chicory and escarole but have not been reported on lettuce. One is chicory X virus, which has a limited host range. It can infect *Gomphrena globosa* L. and two species of *Chenopodium*, in addition to *Cichorium* spp. (Gallitelli and Di Franco, 1982). This virus causes irregular yellow discoloration on the leaves. The other is chicory yellow mottle virus, which is found in southern Italy (Vovlas *et al.*, 1971). It causes a yellow leaf mottle and chlorotic rings and spots.

Turnip mosaic virus is found worldwide and is an occasionally serious problem on both lettuce and endive. However, chicory cultivars tested have been found to be resistant (Provvidenti *et al.*, 1979). Since chicory and endive are cross-compatible, chicory could be a source of resistance for endive and escarole.

Stanghellini and Kronland (1982) reported a fungal root rot of chicory in Arizona, caused by both *Phytophthora cryptogea* Pethyb. and Laff. and *Phymatotrichum omnivorum* (Shear) Dugg.

INSECTS, WEEDS AND OTHER PESTS AND THEIR CONTROL

INSECT PESTS OF LETTUCE

The insect species that are economically important on lettuce and its salad relatives are primarily members of four orders: the Homoptera, the Lepidoptera, the Diptera and the Hymenoptera. Insects can be important in several ways. They may cause damage indirectly by their vectoring capacity, transmitting viruses and virus-like entities that cause disease. These include the aphids and whiteflies (Homoptera), which transmit viruses, and leafhoppers (Homoptera), which transmit phytoplasma-like organisms. Many insects cause damage by feeding: lepidopterous larvae and beetles chew on foliage and stems, aphids and whiteflies suck plant juices, dipterous larvae mine the foliage, wireworms feed on roots and stems. Finally, insects may lower the value of a crop by causing unsightliness. This might include: (i) the presence of aphid or whitefly colonies, together with cast skins and detritus, on the edible portion; (ii) the presence of leafminer mines and larvae on lower leaves; and (iii) holes and ragged edges on leaves resulting from chewing. In addition, some of the insecticides used may be phytotoxic and, if used improperly, toxic to humans and/or detrimental to the environment.

Insects may be controlled by use of insecticides alone or by some level of integrated pest management (IPM). The latter includes: (i) the use of various combinations of resistant cultivars; (ii) biological control (predators and parasitoids); (iii) microbial control (parasitism by microorganisms); (iv) biochemical control (mass trapping or disruption with pheromones); (v) trap crops; (vi) adjusted planting schedules; (vii) elimination of alternate hosts; (viii) field monitoring; and (ix) other methods of avoiding insect crop damage. Treatment with insecticides is by far the most common means of control, although the use of other methods is increasing.

As a result of the heavy use of insecticides to control insects, it is common for various insect species to develop tolerance to the chemicals. This may require increased dosages or more frequent application. Insects may become so resistant to the material that it is no longer useful, and a substitute must be found. Tolerance in two of the groups discussed below, the whiteflies and lepidopteran worms, has been closely studied in recent years.

Homoptera

Aphids
This order includes a number of aphid species, which may inflict harm by transmitting viruses that cause disease, by sucking plant juices, by causing unsightliness or by a combination of these. There are five major species that feed and colonize on above-ground portions and one that is confined to the roots.

The green peach aphid (*Myzus persicae* Sulzer) is the most common aphid found on lettuce. The aphid is pale green, pink, brown or yellowish. It has no waxy covering. Both winged migrant and wingless adult forms occur. Green peach aphid females produce live offspring without fertilization. On lettuce, green peach aphids begin colonizing on lower leaves and may move to the younger ones. They can cause harm to the lettuce plant in two ways, by feeding or as virus vectors. Feeding, by sucking plant juices, can stunt the growth of plants if the colony is sufficiently large.

This aphid is the most important vector of the virus that causes lettuce mosaic. It is also the vector for beet western yellows and turnip mosaic viruses. In most areas, populations build in the early spring and late summer, followed by massive flights, which result in disease expansion from field to field and plant to plant within a field.

The feeding damage caused by the aphid can be controlled in several ways. Most common is the use of various insecticides. During early lettuce growth, insecticide should be applied when the population size begins to increase. During the period before heading starts, moderate numbers can be tolerated without spraying. Plants with large numbers of aphids should be treated to prevent movement of the insects into the centre portion of the plant, where they are more difficult to control. Predators and parasitoids may restrict the size of the populations, but in most cases are not sufficiently effective to be counted on for control. Control of the size of the population in these ways is only partial and is not sufficient to prevent the vectoring of viruses.

Controlling the insect as vector requires other methods. This can be done by the planting of virus-resistant cultivars (see Chapters 2 and 7). The alternative is the use of cultivars resistant to the aphid. In Holland, Eenink and Dieleman (1977) screened 645 accessions in a greenhouse macrotest and identified 14 accessions with potential, partial resistance. These 14 accessions were then compared with five susceptible accessions in a microtest using leaf cages. The microtest essentially confirmed the results of the macrotest. Partial resistance is measured in terms of reduced insect numbers and biomass.

Reduced numbers are a function of increased larval mortality and decreased fecundity. It was easier to discriminate between partial resistance and susceptibility by comparing numbers on older rather than younger plants (Eenink and Dieleman, 1980). Reinink *et al.* (1988) intercrossed six lettuce cultivars with partial resistance and found that it was possible to select to a higher level of resistance. This was probably due to recombination of different genes from different sources. Later, Reinink and Dieleman (1989a) showed that lines with a dominant gene for resistance to the lettuce aphid, *Nazonovia ribis nigri* (Mosley), also had partial resistance to *M. persicae*, probably acting in a polygenic manner.

Breeding for resistance to the green peach aphid is done primarily in Holland, by several of the commercial seed companies, using materials supplied by public researchers.

International shipment of lettuce can be hampered by the presence of live aphids, especially green peach aphids, in lettuce heads. Fumigants used at the receiving point, such as methyl bromide, are dangerous and can also be phytotoxic. Two other materials have been proposed for use at shipping point by vacuum fumigation: acetaldehyde and ethyl formate (Aharoni *et al.*, 1979; Stewart and Aharoni, 1983).

The lettuce aphid (*N. ribis nigri*) is considered the most important leaf aphid on lettuce in Great Britain and Holland. It is found in both the eastern and western USA, but has not been documented as a serious pest of lettuce (Forbes and MacKenzie, 1982). However, they reported a serious outbreak on lettuce in British Columbia, Canada, in 1981 and it is considered a serious pest there.

It is a medium-sized, olive-green aphid. Its primary hosts are *Ribes* spp. and secondary hosts are members of the *Compositae*. It transmits cucumber mosaic virus and beet western yellows virus, but not lettuce mosaic virus. It is more important as a colonizing and feeding species on lettuce, since the aphids tend to get inside the head and multiply there, where they cannot be effectively controlled by insecticides.

In Holland, substantial research has been carried out on resistance to the lettuce aphid. Several accessions of *Lactuca virosa* were found to be highly resistant. Resistance was measured in terms of number of larvae produced. The resistance was transferred to cultivated lettuce by using *Lactuca serriola* as a bridge species between *Lactuca sativa* and *L. virosa* (Eenink *et al.*, 1982a). Resistance is conferred by a single gene, which is incompletely dominant for resistance (Eenink *et al.*, 1982b). This gene also confers partial resistance to *M. persicae*, which may be supplemented by polygenic resistance.

Other foliar aphids which are of some importance on lettuce are the potato aphid (*Macrosiphum euphorbiae* Thos.) and the lettuce brown aphid (*Uroleucon sonchi* L.). The former is more frequently found on lettuce than the latter (Reinink and Dieleman, 1989b).

In a screening test, a low level of resistance to the potato aphid was found among six breeding lines and one cultivar (Reinink *et al.*, 1989). Furthermore, upon screening 90 more lettuce genotypes, considerably greater variation in resistance was found, to both the potato aphid and the lettuce brown aphid (Reinink and Dieleman, 1989b). Partial resistance, primarily of the additive type, was identified in the cultivar Charan against both aphids, and in the cultivar Marbello against the potato aphid only (Reinink *et al.*, 1995).

The lettuce-seed stem aphid (*Macrosiphum barri* Essig) can be a pest on lettuce-seed crops (Carlson, 1959). Treatment with insecticides is the only known control.

The lettuce-root aphid (*Pemphigus bursarius* L.) is a serious pest of lettuce in a number of countries, including Germany, Great Britain, Canada and the USA, and probably in other countries as well. It feeds on lettuce roots, causing wilting of the outer leaves; heads remain small and soft (Fig. 8.1). If the

Fig. 8.1. Effect of lettuce-root aphid. Wilted leaves of susceptible crisphead lettuce on left, upright leaves of resistant cv. Avoncrisp at right.

infestation is heavy, plants may die. Infestations are heaviest in late summer and early autumn.

The primary host group of the aphid includes several species of poplar, but the principal one is the Lombardy poplar (*Populus nigra* var. *italica*) (Dunn, 1959). Winged females fly to the trees in the autumn, mate and lay eggs in the bark crevices. These overwinter, hatch in the spring, multiply by 100-fold or more, mature and form wings and fly to lettuce. They feed and multiply on the roots of lettuce and certain other *Asteraceae*. Large numbers of waxy particles are formed around the insect and in the nearby soil. In late summer, winged forms reappear and migrate back to the poplars. In dry years, the insect can overwinter in the field in the wingless form.

Root aphids can be controlled in several ways. Use of systemic insecticides provides some control when applied before planting. An isolate of the entomopathogenic fungus *Metarhizium anisopliae* was found to be pathogenic to lettuce-root aphids (Chandler, 1997). In the Salinas Valley, at two different periods, in the 1960s and early 1990s, attempts were made to control the aphid by removing Lombardy poplar trees in the lettuce-growing areas. This did not work well in the first experiment, but appeared to be more effective the second time, probably because the clean-up was more extensive.

The most effective control is by use of resistant cultivars. Resistance was at first thought to be controlled by extranuclear factors (Dunn, 1974). However, a subsequent study by Ellis *et al.* (1994) showed that resistance was nuclear and that it was governed by a single dominant allele. This resistance is of a high level and virtually eliminates colonization on roots. The gene is linked closely to *Dm-6*, which confers downy mildew resistance, and the latter can be used as a selection tool for root aphid resistance (Crute and Dunn, 1980). The English cultivars Avoncrisp (crisphead) and Avondefiance (butterhead) are both resistant (Table 8.1); this resistance is traced back to the *Dm-6* resistance source, Plant Introduction (PI) 91532, an accession of *L. serriola*. The American crisphead Empire and the leaf type Grand Rapids are also resistant, but the relationship among resistances has not been determined. Breeding programmes to incorporate root aphid resistance into additional cultivars are in progress in the USA, the UK and Holland.

Whiteflies

Another important group in the Homoptera includes the whiteflies. Whiteflies may cause damage to plants in several ways: as vectors of viruses, by feeding damage and by deposits of honeydew. They thrive in high temperatures and consequently are most common in tropical and subtropical areas. The whiteflies that are troublesome on lettuce include the greenhouse whitefly (*Trialeurodes vaporariorum* Westwood) and two forms of *Bemisia*. They act principally as virus vectors.

The greenhouse whitefly is very commonly found in covered structures, as well as in the field in warm climates. It is the sole transmitter of beet

Table 8.1. Comparison of population development of lettuce-root aphids on susceptible (Webb's Wonderful) and resistant (Avoncrisp) cultivars. (From Ellis *et al.*, 1994.)

Cultivar	Inoculum level (aphids per plant)	Adults per plant		
		14 days	18 days	22 days
Webb's Wonderful	2	12.2	10.0	74.8
	4	24.0	12.5	47.7
	6	17.8	11.3	47.2
Avoncrisp	2	0.0	0.5	0.0
	4	0.3	0.0	0.0
	6	1.3	0.3	0.0

pseudo-yellows on lettuce. It has a large host range and transmits virus to many of the hosts, including cucumber and other cucurbits, sugar beet, spinach, carrot and flax (Duffus, 1965). The virus is semipersistent in the insect host.

The identification of the two forms of *Bemisia* is under dispute (see also Chapter 7). The sweet potato whitefly (*Bemisia tabaci* Gennadius) was described on tobacco in Greece in 1889. It was first identified in Florida in 1894 and in California in the 1920s. It became a severe problem on cotton in Africa in the 1950s. This whitefly attacks over 350 plant species, including lettuce, sugar beet, tomato, cucurbits and cotton, as well as a number of ornamental and weed species. It is also known to transmit viruses in Africa, Asia and other parts of the western hemisphere. In the summer of 1981, there was an enormous increase of whiteflies on melon and cotton in the California and Arizona deserts, followed by a similar population increase on lettuce. Yellowing, necrosis and stunting that occurred on lettuce were diagnosed as symptoms of a new virus disease transmitted by the sweet potato whitefly (Duffus *et al.*, 1986). This disease was named lettuce infectious yellows and became a serious problem, particularly in the early crop of lettuce, which begins growth during the early autumn period when the number of whiteflies remained high. In 1991, a new form of the whitefly was observed, which quickly replaced the original type. This form was considerably more fecund than the original, but considerably less effective in transmitting infectious yellows virus. Two consequences developed: the new form replaced the old and infectious yellows virtually disappeared. However, shortly thereafter, a new yellowing virus appeared, which was named lettuce chlorosis and was vectored by the new whitefly (Duffus *et al.*, 1996).

The similarities and differences between the forms led to two approaches to classification of them. One group maintained that the new form was a different strain (strain B) of the original form (strain A). There were strong similarities in ribonucleic acid (RNA) and deoxyribonucleic acid (DNA) characteristics and, apparently, crosses occurred between them (Campbell *et al.*, 1993). The other group could not obtain crosses in an experiment run under somewhat different cicumstances. They also found other genomic and behavioural differences and therefore named it as a new species (*Bemisia argentifolii*) (Perring *et al.*, 1993). Both forms are almost indistinguishable morphologically.

Bemisia whiteflies are primarily harmful on lettuce as virus vectors, but their feeding may also have other effects, including yellowing and stem blanching (Costa *et al.*, 1993). There appears to be a direct relationship between whitefly density and severity of weight loss. However, normal growth resumes after removal of the flies. These results suggest the action of a toxin.

The sizes of whitefly populations in the Imperial Valley change during the year, increasing in the summer and decreasing substantially during the autumn season. However, small enclaves remain on some hosts through the winter, which are sufficient to start the cycle again in the spring (Coudriet *et al.*, 1986).

Bemisia tabaci has been considered the major virus vector among the few *Bemisia* species known to be vectors. Its relatively sudden development in the 1950s in Africa has been attributed to the increased effectiveness of organic insecticides in destroying natural enemies. Similarly, in the southern tier of states in the USA, increased use of organophosphate and pyrethroid insecticides led to development of resistance and the consequent population explosion in 1981 (Prabhaker *et al.*, 1985). Resistance was detected in field populations collected in the Imperial and Coachella Valleys.

At the present time, the whiteflies in the south-west US deserts are being controlled by application of a systemic insecticide, imidacloprid, under the row before seeding. It is not known how long this control will be effective. The greenhouse whitefly can be controlled in greenhouses with the same chemical or with the parasite *Encarsia formosa* Gah. No sources of resistance to the insects are known.

The aster leafhopper (*Macrosteles quadrilineatus* Forbes) represents the third group in the Homoptera that is economically important for lettuce. This leafhopper is the vector that transmits aster yellows (see also Chapter 7). The host range of both the vector and the disease is quite broad, although lettuce appears to be a preferred host. Control is obtained by repeated insecticide applications; this is usually effective, reducing the numbers of leafhoppers and the incidence of the disease. Reflective aluminium mulch can also be effective (Zalom, 1981).

Lepidoptera

Several species of Lepidoptera can cause serious damage to lettuce. These include the cabbage looper (*Trichoplusia ni* Hubner) (Fig. 8.2), the beet armyworm (*Spodoptera exigua* Hubner), the maize earworm (*Helicoverpa zea* Boddie) and the tobacco budworm (*Heliothis virescens* F.). Damage is caused by larval feeding on the leaf and petiole tissues.

The cabbage looper has been a pest of lettuce in the desert south-west of the USA for many years but little had been published until McKinney (1944) provided data from a 5-year study. The looper feeds on cruciferous species, cucurbit species, tomato, potato, several ornamentals and a number of wild species, as well as lettuce. It is particularly important as a pest in the autumn plantings of lettuce and least damaging on the overwintered spring crops. The larvae feed on the leaves. At young stages of lettuce growth, the larvae can consume the whole plant, or destroy the growing point, which prevents

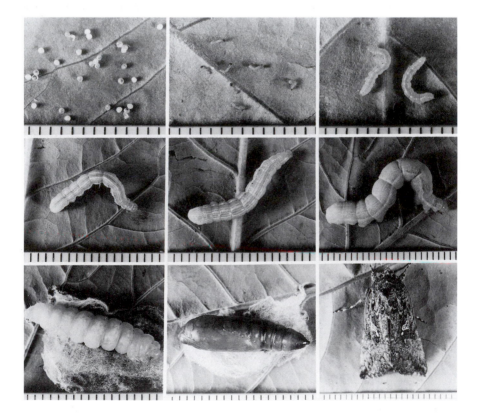

Fig. 8.2. Life stages of cabbage looper from egg to adult moth (courtesy of A. Kishaba).

development of the head. During the heading stage, the insect feeds on the outer leaves, making large holes similar to those made by hail, and occasionally penetrates to the interior. Young plants between seedling and heading stages can tolerate feeding without yield or quality loss.

Moths are about 2 cm long, grey-brown on top, sandy underneath, and have a silver marking on the forewing (Whitaker *et al.*, 1974b). They are primarily nocturnal. Eggs are laid singly on the undersurface of the leaf. They hatch into pale green larvae in about 3 days and mature as dark green individuals 2–2.5 cm long.

Insecticides are the primary means of control of the cabbage looper. There are also a number of parasitoids, predators and microbial agents, but these are not usually effective under natural field conditions. They include a nuclear polyhedrosis virus, species of parasitic wasps and a tachinid fly (*Roria ruralis* Fallen) (Brubaker, 1968). Control by means of resistance is possible. It has been shown that several PI lines of *Lactuca saligna* are resistant (Whitaker *et al.*, 1974a). They showed that the principal mechanism of control was antibiosis, and that there was also a non-preference factor when the larvae were given a choice. Later, ovipositional antixenosis was demonstrated on *L. serriola* and shown to be independent of plant type (Kishaba *et al.*, 1980) (Table 8.2). A breeding programme to transfer resistance to crisphead lettuce was started but discontinued after several years.

The beet armyworm feeds on several crops, including sugar beet, tomato, cotton and alfalfa. It is destructive on lettuce, particularly in the desert southwest of the USA. The moth is small and has mottled brown or grey front wings and lighter grey hind wings. Eggs are laid in clumps and are covered with white hair-like scales. Larvae are usually olive-green with thin, wavy, light-coloured stripes along the sides. The insect is most prevalent in late autumn.

The beet armyworm is attacked by a number of parasites, including wasps and a tachinid fly, as well as viruses. However, control is usually obtained by spraying with an insecticide, preferably before dawn or after sunset when the larvae are most active. No resistance has been identified.

Brewer and Trumble (1989) developed a technique for monitoring insecticide resistance in populations of the beet armyworm. In subsequent work (Brewer *et al.*, 1990) found resistance to methomyl and two other insecticides. Resistance can be detected in both adults and larvae.

Table 8.2. Mean number of cabbage looper eggs on resistant and susceptible parents and their reciprocal F_1 hybrids. (From Kishaba *et al.*, 1980.)

Entry	Generation	No. of plants	Eggs per plant
PI 274372 (*L. serriola*)	P_1 (resis.)	25	17
54671 (*L. sativa*)	P_2 (suscep.)	15	213
34171	F_1 ($P_1 \times P_2$)	7	54
34172	F_1 ($P_2 \times P_1$)	6	28

The maize earworm is common all over California, particularly in the coastal areas. It can wipe out young stands and can also bore into mature heads, rendering them unmarketable. The similar tobacco budworm is more common in the south-western desert areas. The moth of the maize earworm is light tan in colour. It lays single eggs, which are white at first and develop a dark red ring later. Larvae vary in colour, and may be red and blue, red, light green, dark brown and black.

The larvae can decimate seedling stands. They bore into the heads of more mature plants. Natural enemies, including viruses, wasps and bugs, can reduce populations. However, control by insecticides is the usual method.

Other lepidopterous insects can also cause problems on lettuce. Cutworms (four species) are particularly troublesome on young seedlings; they feed on the crown area, killing the foliar portion. Three other species of armyworms also feed on lettuce. The saltmarsh caterpillar (*Estigmene acrea* Drury) may migrate from other crops and cause damage to lettuce.

Diptera

Several species of leafminer can cause damage to lettuce. Among the most serious pests in this group is the pea leafminer (*Liriomyza huidobrensis* Blanchard). It has a history of causing damage to vegetable and ornamental crops in South America and in California. It is a relatively recently reported pest of lettuce. In the Salinas Valley of California, this leafminer causes serious damage to spinach and celery as well as to lettuce. Damage is caused by the larva, which hatches from the egg laid by the mature female, penetrates the leaf surface and burrows through the blade of the leaf. Lower leaves are the first ones attacked. The mines created by the insect are very unsightly and the affected leaves must be trimmed off. When the infestation is severe, more leaves than desirable must be removed, lowering the value of the crop. Two related species, *Liriomyza trifolii* Burgess and *Liriomyza sativae* Blanchard, are particularly troublesome in desert crops of the US south-west. Another leafminer species that is damaging on lettuce is the chrysanthemum leafminer (*Chromatomyia syngenesiae* Hardy).

Hymenoptera

Members of the Hymenoptera that may be pests on lettuce include flea beetles (*Phyllotreta* spp.), darkling beetles (*Blapstinus* spp.), field crickets (*Gryllus* spp.) and wireworms, the larvae of click beetles (various species). The first three are primarily pests of the seedling stand. In the south-west deserts of the USA, they attack the early autumn plantings. Wireworms live in the soil and feed on roots or stems. They are slender, cylindrical and yellowish.

INSECT PESTS OF ENDIVE AND CHICORY

Many of the aphids and lepidopterous worms that attack lettuce also attack endive and chicory.

The maggots of two dipterous flies, *Ophiomyia* spp. and *Napomyza* spp., are pests of witloof chicory. The adult females lay their eggs on the leaves, and produce maggots that burrow in the petiole near the crown. During forcing, they feed on emerging chicon leaves.

Weed control in witloof chicory is primarily with selective herbicides (Bulcke *et al.*, 1986). This is largely successful, but related species of *Asteraceae* are difficult to control. This difficulty can be partially overcome with: (i) the use of soil sterilants the previous autumn; (ii) postemergence low-level treatment with less selective materials in an emergency situation; or (iii) transplanting the chicory in paper pots, forming a temporary physical barrier.

Chlorosulphuron is a highly effective, broad-spectrum, weed-control agent, with low toxicity and requiring only small amounts for control. A single gene for resistance has been identified in witloof chicory, which may allow use of the material on chicory plantings (Lavigne *et al.*, 1994).

The ecological concern for use of herbicide-resistant crops has been discussed for lettuce. A study on the use of the above-mentioned chloro-sulphuron on witloof chicory has been described by Lavigne *et al.* (1995). Two near-isogenic lines of chicory were compared for various traits in field plots. No significant difference was found between them for any vegetative or repro-ductive trait. This suggested that the mutation causing resistance does not have a deleterious effect on plant fitness. If so, it implies that, if the resistance can be transferred to wild chicory plants, there would be no selection against the latter. Therefore they would survive despite the herbicide treatment.

INSECTICIDE USE

The desirability of insecticide use on crop plants is one of the important agricultural issues of recent years. One specific issue involves the amount of active ingredient used. This is a function of amount used per treatment and the frequency of treatment. The standard method of treatment is the application of materials on a regular schedule, regardless of the density of the insect or the effect on the plant (Johnson *et al.*, 1984). A reduced number of applications, based on insect density per plant is often sufficient to control the insect and to maintain the crop. Regular spraying increases the total amount of insecticide applied and may also cause yield reduction. Protection at the early growth stage and at the time of heading is often sufficient to achieve economic control without causing phytotoxicity. Yield reduction varies depending upon the specific material used: methyl parathion caused greater yield reduction than methomyl.

WEEDS

Weeds affect lettuce crops adversely in a relatively passive manner, by growing in the same field and absorbing sunlight, water and nutrients that are needed for the crop. Weeds are good competitors, especially if they emerge at the same time as the crop, or earlier, and have a growth rate as high as or higher than that of the lettuce. Weeds can also serve as hosts for disease organisms, insects and other pests that can injure the crop plants. Some of the weeds that can cause problems in lettuce are shown in Table 8.3.

Integrated weed-control programmes employ both mechanical and chemical methods. The first step is usually the application of a preplant herbicide. Relatively few compounds are useful in lettuce. Pronamide controls a wide spectrum of weeds, but not wild lettuce or other *Asteraceae*, such as groundsel, sowthistle and pineapple weed. Weeds that are not controlled or only partially controlled will then germinate and grow. These must be eliminated mechanically with a hoe or by hand within rows at thinning and by cultivation between rows.

Several other practices can be part of an integrated control programme. Plantings made between seasonal crops of weeds, i.e. between summer and winter weeds, usually need less treatment than later or earlier plantings. Alternative crops, in which the control of weeds may be easier than in lettuce, can be used in rotation with lettuce. To prevent weed seeds from starting a new weed crop, old fields, ditch banks, field edges, road verges and fence lines should be disc-harrowed before their seeds are dropped.

Competition from weeds in lettuce has been the subject of a few studies. Roberts *et al.* (1977) showed that annual weeds at densities of 65–315 plants per square metre caused severe loss of marketable yield (Fig. 8.3). However, if the weeds were removed within 3 weeks of emergence, no yield losses occurred, whether or not the field was maintained free of weeds after 3 weeks. The competitive effect of spiny amaranth (*Amaranthus spinosus* L.) was studied on muck soils in Florida (Shrefler *et al.*, 1996). Head firmness and trimmed head weight were reduced by the presence of weeds less than 105 cm from the lettuce plant; untrimmed head weight was reduced at 45 cm or less.

Most herbicide use is predicated on the application of the material on the field before emergence. Therefore the herbicide must be selective: harmful to weed species but not to the crop. No one herbicide of this type is toxic to all weed species. Those that survive must be mechanically removed. Certain herbicides, such as glyphosate, are universally phytotoxic and can be used where no plant survival is desired. It has been suggested, and research has been done, that crop cultivars can be developed that are resistant to these herbicides, allowing the material to be applied while the crop is growing, in order to destroy more weeds and more species of weeds than with a selective preplant type.

It is important to use preplant herbicides with care, as there may be a

Table 8.3. List of weed species commonly associated with lettuce plantings. Common names are for the USA, UK or both. Family codes: Ama, *Amaranthaceae*; Bor, *Boraginaceae*; Gra, *Gramineae*; Cru, *Cruciferae*; Che, *Chenopodiaceae*; Fum, *Fumariaceae*; Ast, *Asteraceae*; Lab, *Labiateae*; Mal, *Malvaceae*; Leg, *Leguminaceae*; Sol, *Solanaceae*; Pol, *Polygonaceae*; Car, *Caryophyllaceae*; Urt, *Urticaceae*; Scr, *Scrophulariaceae*.

Species	Common name	Family
Amaranthus retroflexus L.	Pigweed, common amaranth, redroot	Ama
A. spinosus L.	Spiny amaranth	Ama
Amsinckia intermedia Fisch. and Mey.	Coast fiddleneck	Bor
Avena fatua L.	Wild oat	Gra
Brassica campestris L.	Mustard, wild turnip	Cru
Capsella bursa-pastoris (L.) Moench	Shepherd's purse	Cru
Chenopodium album L.	Lambsquarters, fat hen	Che
C. murale L.	Nettleleaf goosefoot	Che
C. urbicum L.	Goosefoot, upright goosefoot	Che
Digitaria sanguinalis (L.) Scop.	Crabgrass, hairy finger grass	Gra
Echinochloa crusgalli (L.) Beauv.	Barnyardgrass, cockspur	Gra
Eriochloa gracilis (Fourn.) A.S. Hitchc.	South-western cupgrass	Gra
Eruca sativa L.	Rocket, garden rocket	Cru
Fumaria officinalis L.	Fumitory	Fum
Galinsoga spp.	Gallant soldier	Ast
Lactuca serriola L.	Prickly lettuce	Ast
Lamium amplexicaule L.	Henbit, dead nettle	Lab
Malva parviflora L.	Cheeseweed, least mallow	Mal
Matricaria matricarioides (Less) C.L. Porter	Pineapple weed	Ast
M. recutita L.	Scented mayweed	Ast
M. inodora L.	Scentless mayweed	Ast
Melilotus indica (L.) All.	Yellow sweetclover, small melilot, Indian sweetclover	Leg
Phalaris spp.	Canarygrass	Gra
Physalis spp.	Groundcherry, Japanese lantern	Sol
Poa annua L.	Annual bluegrass, annual meadowgrass	Gra
Polygonum spp.	Knotweed	Pol
Portulaca oleracea L.	Common purslane	Por
Rumex crispus L.	Curly dock	Pol
Senecio vulgaris L.	Common groundsel	Ast
Solanum villosum Lam.	Hairy nightshade	Sol
S. nigrum L.	Black nightshade	Sol
Sonchus oleraceus L.	Sowthistle, smooth sowthistle	Ast
Stellaria media (L.) Vill.	Common chickweed	Car
Urtica urens L.	Stinging nettle, small nettle, burning nettle	Urt
Veronica persica Poir	Persian speedwell	Scr

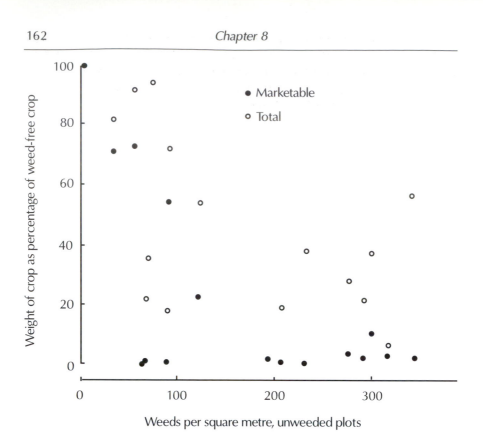

Fig. 8.3. Effect of weed density on marketability of head lettuce. (From Roberts *et al.*, 1977.)

residue problem on a subsequent crop with these as well. Another concern, which would potentially apply to repeated use of herbicides, is that individuals in a weed population that are resistant to the herbicide used will survive and be selected, eventually to replace the susceptible population. That material would no longer control that weed species. The development of sulphonylurea-resistant prickly lettuce has been reported in Idaho (Mallory-Smith *et al.*, 1990).

OTHER PESTS

Animal species other than insects can be damaging to leafy crops by eating the foliage or roots. These include warm-blooded animals, such as birds, deer, rabbits, gophers and mice. A number of bird controls are used. These include noise-makers, coloured strips that move and rustle in the wind, artefacts that look like hawks or other predatory birds and repellents. Repellents can also be used for deer and rabbits. Poisons and traps are used for rodents.

An invertebrate group that can be troublesome includes snails and slugs. These can be controlled with appropriate baits.

MARKETING, ECONOMICS AND FOOD SAFETY

All the activities included in the production of a leafy vegetable crop, from soil preparation, through seeding, cultural practices and protection from pests, to harvesting, are designed to produce a worthwhile product for marketing. The success of marketing is dependent upon several factors not under the control of the grower or of the vagaries of climate and other aspects of the environment. Rather, it is dependent upon the laws of supply and demand, but also upon maintenance of quality during transportation, storage and display on the retail counter.

LETTUCE

The sale of lettuce in a market is dependent upon price and quality. However, the effect of these factors is modified by the fact that, in many countries, lettuce is considered a staple item, and therefore is purchased regularly despite wide variation in prices and quality. Lettuce has to be extremely high-priced or in extremely poor condition before most consumers decide against buying it.

Costs and Price

Several variables enter into the price of lettuce, starting with the costs of producing the crop. These include land preparation, seed and seed coating, planting, water and irrigation, fertilizer and application, pesticides and treatment, cultivation practices, fuel and maintenance. In addition, there are overhead costs, including land rental, taxes and business costs. Finally, there are the costs of harvesting, including labour, vehicles and containers. These are referred to as the farm-gate costs. Add the costs of pallets, hauling from the field and cooling, and the final figure is the cost of production. The so-called break-even price is based upon the cost per container of lettuce. This figure is

164

greatly dependent upon the total yield and is calculated by dividing the total cost per land unit by the number of containers harvested from that unit. The profit to the shipper is the difference between the break-even price and the price actually paid by a receiver. This is the free-on-board (FOB) price. A grower's profit is dependent upon these prices and the specific contract between the grower and shipper.

Other costs then come into play. These include the salaries to salespeople, charges by sales brokers and transportation costs (fuel, cooling, salaries, maintenance and other costs). Add final receiving-point costs (cooling, trimming, wrapping and display). The wholesale price to individual stores is based upon these additional costs; the profit to the retailer is the difference between the sum of these costs and shelf price.

A number of pricing studies have been made in the USA that focus primarily on the lettuce produced in the California industry, by far the largest in the country. Some of these studies have examined the relationship between the price of lettuce paid to the grower and shipper and the price on the retail market. Powers (1995) found that FOB prices and wholesale prices move up and down together, with adjustment taking place in about a week. Retail prices adjust more slowly to FOB and wholesale prices, and the adjustment is slower during a decline than during an increase. A similar study also suggests that retailer power can depress farm prices, especially when supplies are high (Sexton and Zhang, 1995). In a limited competitive situation, where the perishability of the product allows the shipper little room for bargaining, an excess in supply constrains the shipper even more. Therefore, FOB prices are much more volatile than retail prices (Sexton and Zhang, 1996). In an earlier study, Raleigh (1978) found that in the 11-year period between 1967 and 1977, the grower's share of the retail price declined in relation to the wholesaler's and retailer's shares (Fig. 9.1).

Variations in price occur because of supply and demand. Good weather and minimum production problems may create an oversupply. The price offered to the shipper will be lower than if poorer conditions existed and may often be lower than that required to pay the costs of growing and harvesting. Demand may be lower at the destination during periods of severe winter weather. Both supply and demand go up during the summer period, and the price will depend upon the specific ratio between the two. Sometimes the anticipation of a problem that would reduce the supply may increase the price offered.

Quality

Quality maintenance is an essential part of a successful marketing programme. Quality is measured in several ways. One is freedom from disease and insect damage and the unsightliness that accompanies such damage.

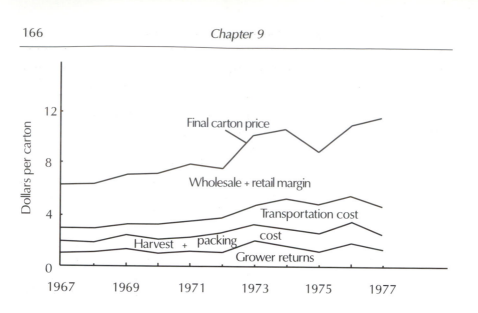

Fig. 9.1. The cumulative effect of intermediate income factors on the final carton price of California lettuce at the terminal market. Average over Chicago and New York markets in USA. (From Raleigh, 1978.)

Other cosmetic attributes are also important, including good colour, relating both to the chlorophyll expression and to anthocyanin expression in cultivars with red colouring. Shape of the head is important, as is head weight, head density and rib structure.

Maintenance of quality may be achieved through internal policies of the producing organization, the demands of the receiver and the consumer and by regulations enforced by governmental agencies. The producer achieves quality by good farming practices, disease and insect control and harvest and postharvest practices designed to keep the product in good condition. The receiver and consumer exert control by inspecting the product and rejecting it if it is unsatisfactory. Supermarket chains usually have defined quality standards. Governmental agencies promulgate regulations defining standards that must be met for a product to qualify for acceptance. In the USA, the federal government defines grades of product in terms of blemishes and other measures of quality. On a more local level, county inspectors assess for tipburn and other signs of damage at the time of harvest and can stop the harvest if the damage exceeds a defined level.

In greenhouse-grown lettuce, in particular, low nitrate nitrogen content in the marketed product is important, most specifically in the European market. Künsch *et al.* (1995), in a lettuce soilless-culture experiment, found that nitrate content could be reduced significantly by reducing the nitrogen input shortly before harvest.

Sensory characteristics are important and can be related to bitterness level caused by sesquiterpene lactones (Price *et al.*, 1990). Among green lettuces tested, leaf, butterhead, cos, crisphead and Batavian were decreasingly bitter. Lactucin glycoside was the main contributor to the bitterness effect (Figs 9.2 and 9.3).

Cut-lettuce products are particularly sensitive to changes in sensory characteristics (Heimdal *et al.*, 1995). Processing followed by storage in normal oxygen (O_2) content atmospheres resulted in browning of cut edges, yellowing and reduction in ascorbic acid content. Moderate vacuum-packaging with reduced O_2 and increased carbon dioxide (CO_2) reduced the severity of these quality problems. Five types of lettuce stored under controlled-atmosphere (CA) conditions (3% O_2, 10% CO_2) showed greater visual quality after 8 days than those stored in air (López-Gálvez *et al.*, 1996). The CA protected all types well for 12 days. The crisphead type benefited most in CA and the butterhead type was least protected. Low concentration of O_2 minimized browning of five cultivars of butterhead lettuce, while low CO_2 content minimized decay (Varoquaux *et al.*, 1996). Both goals could be achieved together by flushing the package with nitrogen (N_2).

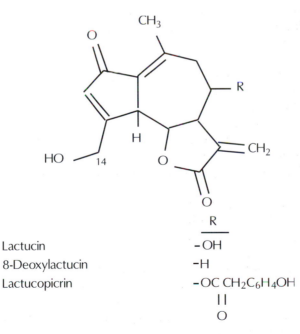

	R
Lactucin	$-OH$
8-Deoxylactucin	$-H$
Lactucopicrin	$-OC\,CH_2C_6H_4OH$ $\underset{O}{\overset{\|\|}{}}$

Fig. 9.2. Chemical structures of the sesquiterpene lactones found in lettuce. (From Price *et al.*, 1990.)

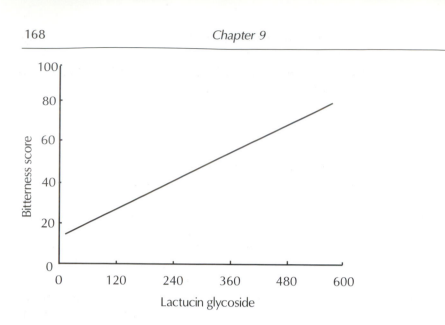

Fig. 9.3. Relation of lactucin glycoside score to bitterness in chicory. (From Price *et al.*, 1990.)

Sales, Advertising and Transportation

Most lettuce is marketed to the consumer through grocery chains. It reaches the chain stores by several methods. It may be sold by the shipper directly to the chain or to a buying organization that supplies several chains. It may be sold through a broker. It may be sold to a terminal market, where buyers for smaller chains and single shops purchase the lettuce.

Bulk lettuce is usually sold directly to grocery and restaurant chains under contract between the shipper and the chain. The demand and the price are inelastic through the life of the contract. Contracts may be 90 or 180 days in length; the inelasticity guarantees a profit to the shipper.

Brand advertising of lettuce, as with most produce items, is mostly limited to the label on the container in which the lettuce is shipped. Several shippers put their labels on the plastic bags in which lettuce is wrapped as well as on the carton containing the wrapped or unwrapped heads. A few advertise through various media. A promotional commission once operated in California as a generic advertising organization, but is no longer in existence.

Lettuce packed in various salad combinations is always under a brand name. This is the fastest-growing method of marketing lettuce, and is very competitive because of product promotion emphasizing brand names.

Nearly all lettuce is shipped in refrigerated lorry trailers. These require 1–4 days to reach the receiving point, depending upon the distance from the shipping point. The goal of refrigeration is to maintain the temperature of the load at 1–2°C, although it may rise higher or sink lower, depending upon season and other factors. Occasionally, lettuce is shipped by rail or by air; lettuce for overseas export usually travels by container ship.

Export

In the USA, lettuce for export constitutes a relatively small portion of the total amount marketed. It is, however, the leading export-vegetable commodity. In 1994/95, the US exports of lettuce were valued at $184 million. About 85% was shipped to Canada, about 5% to Hong Kong and the rest to Mexico, other Asian nations and the UK (Anonymous, 1996) (Table 9.1).

Britain, Germany and Switzerland are the main importing countries in Europe (Hutin, 1995). The UK imports considerably more than it exports, primarily from Spain, France, the Netherlands and the USA. About 1% of the UK crop is exported, mostly to Norway, Sweden and Denmark. The principal exporters in Europe are Spain, Holland, Belgium and France, in that order. France markets half of its export crop to Germany and over 12% to Switzerland. In Australia, approximately 8% of the crop is exported (almost 7500 tonnes in 1992/93), primarily to Singapore, Hong Kong and Japan. Spain, Israel, Morocco, Holland and Belgium export substantial proportions of their production.

ENDIVE

Endive and escarole are minor leafy crops in the USA, as well as in Australia, Israel, Spain, Britain and most other leafy-vegetable-producing countries. In the USA, the two forms of endive are marketed, transported and handled in essentially the same way as leaf lettuces. Endive and escarole are much more widely grown in France, Belgium, Italy and Holland than in any other country. In these countries, they are grown for home consumption and for export.

Table 9.1. Value of exported lettuce from USA to principal markets, fiscal years 1990/91–1994/95. (From Anonymous, 1996.)

Market	Value in $1000				
	1990/91	1991/92	1992/93	1993/94	1994/95
Canada	110,700	99,700	125,200	91,000	157,700
Hong Kong	6,800	7,000	10,100	11,500	9,300
Japan	800	5,600	4,800	6,700	4,600
Mexico	3,500	5,700	7,300	8,800	4,300
Singapore	800	900	1,100	1,700	2,600
Taiwan	1,100	1,100	1,300	1,500	1,800
UK	4,000	2,300	2,300	3,200	1,000

Research on endive quality is limited. Jacques *et al.* (1995) were interested in the implication of population sizes of microbial agents on the subsequent rotting of tissue after harvest. They estimated sizes of bacterial populations of several types on escarole leaves. The population sizes were greater on outer than on inner leaves and were dependent upon environmental interaction factors, as well as exposure to airborne bacteria.

Bitterness is a taste factor in all edible forms of these vegetables, but varies widely among types and cultivars. However, it was found that the bitterness level of endive and chicory cultivars was considerably higher than that of any lettuce (Price *et al.*, 1990). The sesquiterpene lactone content was also correspondingly higher.

CHICORY

Chicory is also a minor leafy crop in the USA and in many other leafy-vegetable-producing countries. The principal form of chicory grown in commercial quantities in the USA is radicchio and it is marketed like butterhead lettuce; very little witloof chicory is grown. Chicory is grown extensively in the same western European countries as endive. However, the witloof type of chicory is not produced in Italy, which grows a great variety of non-forced types.

Witloof chicory is considered a gourmet item in the USA (Corey *et al.*, 1990). It is expensive and used only by a small proportion of the public. In order to reach a greater stage of importance, extensive promotional campaigns and lower prices are necessary. However, imports from Europe, especially Belgium, are increasing.

There has been substantial expansion in European witloof chicory production in recent years. Consumption has increased in areas where previously it had been quite low, such as Germany and southern France. Export has increased in importance; about 40% of Belgian production and 20–25% of Dutch production is for export.

Sale by auction is common in western Europe. A price is set by the auctioneer based upon the condition of the product. The buyers in the group may then bid silently for the unit of product; the first bidder at that price buys the unit. If there is no bidder at that price, the price is lowered by a fixed amount and the bidding starts again. This procedure continues until the unit is sold.

Chicory quality characteristics are similar to those of lettuce. However, there is an additional factor pertaining to the witloof type: chicons should be white to yellowish white. Exposure to light, such as on a supermarket shelf, will cause the development of green colour, which is undesirable since it is considered unsightly and increases the level of bitterness.

SAFETY ISSUES

In recent years, an interest in safety has arisen in the marketing of fresh fruits and vegetables. It relates to the possible association of pathogenic organisms with the product (Beuchat, 1996) (Table 9.2). The basis for this interest is the increasing use of fresh and lightly processed produce (known as grade 4 products in Europe), along with the movement of these products from one region to another with different health and safety standards. In addition, handling, processing, packaging and distribution within regions is of interest for identifying and controlling possible hazards. Hazard analysis critical control point (HACCP) programmes are being initiated to attempt to minimize chances of illness (Willocx *et al.*, 1993) (Table 9.3). These include monitoring sanitation, raw product, packaging, storage and distribution procedures for health hazards, including pathogenic organisms, chemicals and physical entities.

The exposure of cut surfaces of minimally processed packaged lettuce may lead to invasion by microorganisms. Some of these may be harmful. Of special concern is *Escherichia coli* 0157:H7, a rare form of the common bacterium, which causes severe gastrointestinal disorder (Diaz and Hotchkiss, 1996). A study of the growth of *E. coli* 0157:H7 on cut lettuce under various storage conditions showed that storage in air at 22°C led to a higher rate of growth of the organism than at 13°C. However, storage at 13°C, under a modified atmosphere that permitted the longest shelf-life, was conducive to growth of the organism comparable to storage at 22°C under ambient conditions.

Table 9.2. Sources of pathogenic organisms of fresh produce and conditions that influence their survival and growth. (From Beuchat, 1996.)

Preharvest	
Faeces	Dust
Soil	Wild and domestic animals
Irrigation water	Human handling
Green or inadequately composted manure	
Postharvest	
Faeces	Ice
Human handling (workers, consumers)	Transport vehicles
Harvest equipment	Improper storage temperature
Transport containers	Improper packaging
Wild and domestic animals	Cross-contamination from other foods
Dust	Improper display temperatures
Wash and rinse water	Improper handling after wholesale or retail purchase
Processing equipment	

Table 9.3. The seven basic principles of the hazard analysis critical control point (HACCP) system. (From Willocx *et al.*, 1993.)

1. Identification of the potential hazard(s) (hazard analysis)
2. Determination of the points/procedures/operational steps critical to the identified hazard(s) (critical control point(s) (CCP))
3. Establishment of critical level(s) and tolerance(s) which must be met to ensure each CCP is under control
4. Establishment and implementation of a monitoring system to ensure control of the CCP
5. Identification and execution of the corrective action to be taken if a deviation occurs at a CCP
6. Establishment of a documentation system, including all procedures and records appropriate to all the principles and their application
7. Verification to confirm (by supplementary procedures and tests) that the HACCP system is working effectively

Hagenmaier and Baker (1997) showed that low-level ionizing irradiation of cut and packaged iceberg lettuce, combined with chlorination, minimized microbial populations on the product.

A study of the impact of an HACCP system on safety and quality of grade 4 products has been made, using witloof chicory (Belgian endive) as a model product (Willocx *et al.*, 1993). The effect of temperature was measured on oxidation of tissues and growth of test organisms, *Pseudomonas marginalis* and *Lactobacillus plantarum*. They concluded that it is possible to predict deterioration under fluctuating temperatures based upon parameters developed under controlled conditions.

Magnuson *et al.* (1990) sampled fresh and cut lettuce to assess the number and diversity of microorganisms found (Table 9.4). They found that the number and types of bacteria and yeasts on fresh lettuce differed from those on processed lettuce, whether in sealed or unsealed packages.

King *et al.* (1991) found that the count of microbial flora increased even at low-temperature storage. Package atmosphere composition and pH also

Table 9.4. Bacterial and yeast genera found most commonly in fresh and packaged lettuce. (From Magnuson *et al.*, 1990.)

Isolates	Fresh	Packaged
Bacteria	*Erwinia*	*Pseudomonas*
	Pseudomonas	*Erwinia*
	Serratia	
Yeasts	*Candida*	*Candida*
	Cryptococcus	*Cryptococcus*
	Pichia	*Pichia*
	Trichosporon	*Torulaspora*

changed with time. The degree of change was also dependent upon the wrapping material. Sealed commercial packages retarded growth of bacteria longer, but had less effect on yeast growth.

Carlin *et al.* (1996) found that a storage atmosphere of 10% each of O_2 and CO_2 was most conducive to visual quality of shredded escarole and had minimum effect on growth of aerobic bacteria and *Listeria monocytogenes*.

REFERENCES

Abawi, G.S. and Grogan, R.G. (1979) Epidemiology of diseases caused by *Sclerotinia* species. *Phytopathology* 69, 899–904.

Abawi, G.S., Grogan, R.G. and Duniway, J.M. (1985) Effect of water potential on survival of sclerotia of *Sclerotinia minor* in two California soils. *Phytopathology* 75, 217–221.

Adams, P.B. (1987) Effects of soil temperature, moisture, and depth on survival and activity of *Sclerotinia minor*, *Sclerotium cepivorum*, and *Sporidesmium sclerotivorum*. *Plant Disease* 71, 170–174.

Adams, P.B. and Ayers, W.A. (1982) Biological control of *Sclerotinia* lettuce drop in the field by *Sporidesmium sclerotivorum*. *Phytopathology* 72, 485–488.

Aharoni, Y., Stewart, J.K., Hartsell, P.L. and Young, D.K. (1979) Acetaldehyde – a potential fumigant for control of the green peach aphid on harvested head lettuce. *Journal of Economic Entomology* 72, 493–495.

Aikman, D.P. and Scaife, M.A. (1993) Modelling plant growth under varying environmental conditions in a uniform canopy. *Annals of Botany* 72, 485–492.

Alvarez, J., Datnoff, L.E. and Nagata, R.T. (1992) Crop rotation minimizes losses from corky root in Florida lettuce. *HortScience* 27, 66–68.

Ameziane, R., Cassan, L., Dufosse, C., Rufty, T.W. Jr, and Limami, A.M. (1997) Phosphate availability in combination with nitrate availability affects root yield and chicon yield and quality of Belgian endive. *Plant and Soil* 191, 269–277.

Andersen, E.M. (1946) *Tipburn of Lettuce*. Agricultural Experiment Station Bulletin 829, Cornell University, Ithaca, New York, 14 pp.

Anderson, P.A., Okubara, P.A., Arroyo-Garcia, R., Meyers, B.C. and Michelmore, R.W. (1996) Molecular analysis of irradiation-induced and spontaneous deletion mutants at a disease resistance locus in *Lactuca sativa*. *Molecular and General Genetics* 251, 316–325.

Anonymous (1996) US fresh vegetable exports hit a record in FY 1995. *World Horticultural Trade and US Exports Opportunities* 1, 9–14.

Asirifi, K.N., Morgan, W.C. and Parbery, D.G. (1994) Suppression of sclerotinia soft rot of lettuce with organic soil amendments. *Australian Journal of Experimental Agriculture* 34, 131–136.

Ayers, W.A. and Adams, P.B. (1979) Mycoparasitism of sclerotia of *Sclerotinia* and

Sclerotium species by *Sporidesmium sclerotivorum*. *Canadian Journal of Microbiology* 25, 17–23.

Badila, P., Lauzac, M. and Paulet, P. (1985) The characteristics of light in floral induction *in vitro* of *Cichorium intybus*. The possible role of phytochrome. *Physiologia Plantarum* 65, 305–309.

Ballantyne, A., Stark, R. and Selman, J.D. (1988) Modified atmosphere packaging of shredded lettuce. *International Journal of Food Science and Technology* 23, 267–274.

Bannerot, H. and de Coninck, B. (1976) Breeding 'roodloof' (red Brussel chicory). In: *Proceedings Eucarpia Meeting on Leafy Vegetables*, Institute for Horticultural Plant Breeding, Wageningen, the Netherlands, p. 40.

Bannerot, H., Boulidard, L., Marrou, J. and Duteil, M. (1969) Étude de l'hérédité de la tolérance au virus de la mosaïque de la laitue chez la variété Gallega de Invierno. *Annales Phytopathologie* 1, 219–226.

Barrons, K.C. and Whitaker, T.W. (1943) Great Lakes, a new summer head lettuce adapted to summer conditions. *Michigan Agricultural Research Station Quarterly Bulletin* 25, 1–3.

Barton, L.V. (1966) Effects of temperature and moisture on viability of stored lettuce, onion and tomato seeds. *Contributions to the Boyce Thompson Institute of Plant Research* 23, 285–290.

Bass, L.N. (1970) Prevention of physiological necrosis (red cotyledons) in lettuce seeds (*Lactuca sativa* L.). *Journal of the American Society for Horticultural Science* 95, 550–553.

Bass, L.N. (1973) Controlled atmosphere and seed storage. *Seed Science and Technology* 1, 463–492.

Bassett, M.J. (1975) The role of leaf shape in the inheritance of heading in lettuce (*Lactuca sativa* L.). *Journal of the American Society for Horticultural Science* 100, 104–105.

Beach, W.S. (1921) *The Lettuce 'Drop' due to* Sclerotinia minor. Bulletin No. 165, Pennsylvania State College Agricultural Experiment Station, Pennsylvania, 27 pp.

Bekendam, J., van Pijlen, J.G. and Kraak, H.L. (1987) The effect of priming on the rate and uniformity of germination of endive seed. *Acta Horticulturae* 215, 209–218.

Bellamy, A., Vedel, F. and Bannerot H. (1996) Varietal identification in *Cichorium intybus* L. and determination of genetic purity of F_1 hybrid samples, based on RAPD markers. *Plant Breeding* 115, 128–132.

Bensink, J. (1958) Morphogenetic effects of light intensity and night temperature on the growth of lettuce (*Lactuca sativa* L.) with special reference to the process of heading. *Proceedings of the Royal Netherlands Academy of Science, Series C* 61, 89–100.

Bensink, J. (1961) Heading of lettuce (*Lactuca sativa* L.) as a morphogenetic effect of leaf growth. In: Veenman, N. and Zonen, N.V. *Proceedings of the XVth International Horticultural Congress*. Advances in Horticultural Science 1, Pergamon Press, New York, London, pp. 470–475.

Bensink, J. (1971) On morphogenesis of lettuce leaves in relation to light and temperature. *Mededeling Landbouwhogeschool, Wageningen* 71, 1–93.

Berrie, A.M.M. and Robertson, J. (1973) Growth retardants and the germination of light sensitive lettuce seed. *Physiologia Plantarum* 28, 278–283.

Beuchat, L.R. (1996) Pathogenic organisms associated with fresh produce. *Journal of Food Protection* 59, 204–216.

Bewley, J.D. (1980) Secondary dormancy (skotodormancy) in seeds of lettuce (*Lactuca sativa* cv. Grand Rapids) and its release by light, gibberellic acid and benzyladenine. *Physiologia Plantarum* 49, 277–280.

Biddington, N.L. and Dearman, A.S. (1984) Shoot and root growth of lettuce seedlings following root pruning. *Annals of Botany* 53, 663–668.

Bierhuizen, J.F., Ebbens, J.L., and Koomen, N.C.A. (1973) Effects of temperature and radiation on lettuce growing. *Netherlands Journal of Agricultural Science* 21, 110–116.

Blaauw-Jensen, G. (1981) Differences in the nature of thermodormancy and far-red dormancy in lettuce seeds. *Physiologia Plantarum* 53, 553–557.

Bohn, G.W. and Whitaker, T.W. (1951) *Recently Introduced Varieties of Head Lettuce and Methods Used in Their Development.* Circular No. 881, US Department of Agriculture, Washington, DC, 27 pp.

Bolin, H.R. and Huxsoll, C.C. (1991) Effect of preparation procedures and storage parameters on quality retention of salad-cut lettuce. *Journal of Food Science* 56, 60–67.

Bonnier, F.J.M., Reinink, K. and Groenwold, R. (1994) Genetic analysis of *Lactuca* accessions with new major gene resistance to lettuce downy mildew. *Phytopathology* 84, 462–468.

Borthwick, H.A. and Robbins, W.W. (1928) Lettuce seed and its germination. *Hilgardia* 3, 275–305.

Borthwick, H.A., Hendricks, S.B., Parker, M.W., Toole, E.H. and Toole, V.K. (1952) A reversible photoreaction controlling seed germination. *Proceedings of the National Academy of Science* 38, 662–666.

Borthwick, H.A., Hendricks, S.B., Toole, E.H. and Toole, V.K. (1954) Action of light on lettuce seed germination. *Botanical Gazette* 115, 205–225.

Bos, L. and Huijberts, N. (1990) Screening for resistance to big-vein disease of lettuce (*Lactuca sativa*). *Crop Protection* 9, 446–452.

Boukema, I.W., Hazekamp, T. and van Hintum, T.J.L. (1990) *The CGN Lettuce Collection.* Centre for Genetic Resources, Wageningen, the Netherlands.

Bradford, K.J. (1990) A water relations analysis of seed germination rates. *Plant Physiology* 94, 840–849.

Bremer, A.H. and Grana, J. (1935) Genetische untersuchungen mit salat. II. *Gartenbauwissenshaft* 9, 231–245.

Brewer, M.J. and Trumble, J.T. (1989) Field monitoring for insecticide resistance in beet armyworm (Lepidoptera: Noctuidae). *Journal of Economic Entomology* 52, 1520–1526.

Brewer, M.J., Trumble, J.T., Alvarado-Rodriguez, B. and Chaney, W.E. (1990) Beet armyworm (Lepidoptera: Noctuidae) adult and larval susceptibility to three insecticides in managed habitats and relationship to laboratory selection for resistance. *Journal of Economic Entomology* 83, 2136–2146.

Broadbent, L. (1951) Lettuce mosaic in the field. *Agriculture* 62, 578–582.

Brown, P.R. and Michelmore, R.W. (1988) The genetics of corky root resistance in lettuce. *Phytopathology* 78, 1145–1150.

Brubaker, R.W. (1968) Seasonal occurrence of *Voria ruralis*, a parasite of the cabbage looper, in Arizona, and its behavior and development in laboratory culture. *Journal of Economic Entomology* 61, 306–309.

Bruckart, W.L. and Lorbeer, J.W. (1976) Cucumber mosaic virus in weed hosts near commercial fields of lettuce and celery. *Phytopathology* 66, 253–259.

Budge, S.P., McQuilken, M.P. and Fenlon, J.S. (1995) Use of *Coniothyrium minitans* and *Gliocladium virens* for biological control of *Sclerotinia sclerotiorum* in glasshouse lettuce. *Biological Control* 5, 513–522.

Bukovac, M.J. and Wittwer, S.H. (1958) Reproductive responses of lettuce (*Lactuca sativa* var. Great Lakes) to gibberellin as influenced by seed vernalization, photoperiod, and temperature. *Proceedings of the American Society for Horticultural Science* 71, 407–411.

Bulcke, R., Stryckers, J. and Van Himme, M. (1986) Weed problems in vegetable crops in Belgium. In: *Proceedings of the Meeting of European Community Experts Group, Stuttgart, 28–31 October 1986*, pp. 65–70.

Burdett, A.N. (1972) Ethylene synthesis in lettuce seeds: its physiological significance. *Plant Physiology* 50, 719–722.

Burdett, A.N. and Vidaver, W.E. (1971) Synergistic action of ethylene with gibberellin on red light in germinating lettuce seeds. *Plant Physology* 48, 656–657.

Burkholder, W.H. (1954) Three bacteria pathogenic on head lettuce in New York State. *Phytopathology* 44, 592–596.

Campbell, B.C., Duffus, J.E. and Baumann, P. (1993) Determining whitefly species. *Science* 261, 1333–1335.

Campbell, R.N. (1965) Weeds as reservoir hosts of the lettuce big-vein virus. *Canadian Journal of Botany* 43, 1149.

Campbell, R.N. and Lot, H. (1996) Lettuce ring necrosis, a viruslike disease of lettuce: evidence for transmission by *Olpidium brassicae*. *Plant Disease* 80, 611–615.

Campbell, R.N., Grogan, R.G. and Purcifull, D.E. (1961) Graft transmission of big vein of lettuce. *Virology* 15, 82–85.

Campbell, R.N., Greathead, A.S. and Westerlund, F.V. (1980) Big vein of lettuce: infection and methods of control. *Phytopathology* 70, 741–746.

Caporn, S.J.M. (1989) The effects of oxides of nitrogen and carbon dioxide enrichment on photosynthesis and growth of lettuce (*Lactuca sativa* L.). *New Phytologist* 111, 473–481.

Carette, B. and Laurent, E. (1989) Architecture de la chicorée porte-graine et qualité des semences produties: influence de la densité du peuplement. *Acta Horticulturae* 253, 31–43.

Carlin, F., Nguyen-the, C., Da Silva, A.A. and Cochet, C. (1996) Effects of carbon dioxide on the fate of *Listeria monocytogenes*, of aerobic bacteria and on the development of spoilage in minimally processed fresh endive. *International Journal of Food Microbiology* 32, 159–172.

Carlson, E.C. (1959) Control of *Macrosiphum barri* Essig and its damage to lettuce seed plants. *Journal of Economic Entomology* 52, 411–414.

Carpenter, C.W. (1916) The Rio Grande lettuce disease. *Phytopathology* 6, 303–305.

Ceponis, M.J. (1970) Diseases of California head lettuce on the New York market during the spring and summer months. *Plant Disease Reporter* 54, 964–966.

Chandler, D. (1997) Selection of an isolate of the insect pathogenic fungus *Metarhizium anisosopliae* virulent to the lettuce root aphid, *Pemphigus bursarius*. *Biocontrol Science and Technology* 7, 95–104.

Chang, K.F., Mirza, M. and Hwang, S.F. (1991) Occurrence of lettuce rust in Onoway, Albert in 1989. *Canadian Plant Disease Survey* 71, 17–19.

Chapogas, P.G. and Stokes, D.R. (1964) *Prepackaging Lettuce at Shipping Point.* Marketing Research Report 670, United States Department of Agriculture, Washington, DC, USA, 48 pp.

Chen, X.G., Gastaldi, C., Siddiqi, M.Y. and Glass, A.D.M. (1997) Growth of a lettuce crop at low ambient nutrient concentrations: a strategy designed to limit the potential for eutrophication. *Journal of Plant Nutrition* 20, 1403–1417.

Chen, Y.K. and Chen, M.J. (1994) Lettuce leafroll mosaic – a new lettuce disease caused by caulimovirus-like agent in Taiwan. *Plant Pathology Bulletin* 3, 209–215.

Cho, J.J., Mitchell, W.C., Mau, R.F.L. and Sakimura, K. (1987) Epidemiology of tomato spotted wilt virus disease on crisphead lettuce in Hawaii. *Plant Disease* 71, 505–508.

Cho, J.J., Mau, R.F.L., German, T.L., Hartmann, R.W., Yudin, L.S., Gonsalves, D. and Providenti, R. (1989) A multidisciplinary approach to management of tomato spotted wilt virus in Hawaii. *Plant Disease* 73, 375–383.

Christie, S.R., Edwardson, J.R. and Zettler, F.W. (1968) Characterization and electron microscopy of a virus isolated from *Bidens* and *Lepidium. Plant Disease Reporter* 52, 763–768.

Cichan, M. (1983) Self fertility in wild populations of *Cichorium intybus* L. *Bulletin of the Torrey Botanical Club* 110, 316–323.

Collier, G.F. and Tibbitts, T.W. (1982) Tipburn of lettuce. *Horticultural Reviews* 4, 49–65.

Coons, J.M., Kuehl, R.O. and Simons, N.R. (1990) Tolerance of ten lettuce cultivars to high temperature combined with NaCl during germination. *Journal of the American Society for Horticultural Science* 115, 1004–1007.

Coppens d'Eeckenbrugge, G., van Herck, J.C. and Dutilleul, P. (1989) A study of fructose yield components in chicory. *Plant Breeding* 102, 296–301.

Corbineau, F. and Come, D. (1990) Germinability and quality of *Cichorium intybus* L. seeds. *Acta Horticulturae* 267, 183–189.

Corey, K.A., Marchant, D.J. and Whitney, L.F. (1990) Witloof chicory: A new vegetable crop in the United States. In: Janick, J. and Simon, J.E. (eds) *Advances in New Crops.* Timber Press, Portland, pp. 414–418.

Costa, A.S. and Duffus, J.E. (1958) Observations on lettuce mosaic in California. *Plant Disease Reporter* 42, 583–586.

Costa, H.S., Ullman, D.E., Johnson, M.W. and Tabashnik, B.E. (1993) Association between *Bemisia tabaci* density and reduced growth, yellowing, and stem blanching of lettuce and kai choy. *Plant Disease* 77, 969–972.

Costigan, P.A. (1986) The effects of soil type on the early growth of lettuce. *Journal of Agricultural Science, Cambridge* 107, 1–8.

Couch, H.B. (1955) Studies on seed transmission of lettuce mosaic virus. *Phytopathology* 45, 63–70.

Couch, H.B. and Gold, A.H. (1954) Rod-shaped particles associated with lettuce mosaic. *Phytopathology* 44, 715–717.

Couch, H.B. and Grogan, R.G. (1955) Etiology of lettuce anthracnose and host range of the pathogen. *Phytopathology* 45, 375–380.

Coudriet, D.L., Meyerdirk, D.E., Prabhaker, N. and Kishaba, A.N. (1986) Bionomics of sweetpotato whitefly (Homoptera: Aleyrodidae) on weed hosts in the Imperial Valley, California. *Environmental Entomology* 15, 1179–1183.

Cox, E.F. and Dearman, A.S. (1981) The effect of trickle irrigation, misting, and row position on the incidence of tipburn of field lettuce. *Scientia Horticulturae* 15, 101–106.

Cox, E.F. and McKee, J.M.T. (1976) A comparison of tipburn susceptibility in lettuce under field and glasshouse conditions. *Journal of Horticultural Science* 51, 117–122.

Crescenzi, A., Nuzzaci, M., De Stradis, A., Comes, S. and Piazzolla, P. (1996) Characterization of a new virus from escarole. *Annals of Applied Biology* 128, 65–75.

Crisp, P., Collier, G.F. and Thomas, T.H. (1976) The effect of boron on tipburn and auxin activity in lettuce. *Scientia Horticulturae* 5, 215–226.

Crute, I.R. (1984) The integrated use of genetic and chemical methods for control of lettuce downy mildew (*Bremia lactucae* Regel). *Crop Protection* 3, 223–241.

Crute, I.R. and Dickinson, C.H. (1976) The behaviour of *Bremia lactucae* on cultivars of *Lactuca sativa* and on other composites. *Annals of Applied Biology* 82, 433–450.

Crute, I.R. and Dunn, J.A. (1980) An association between resistance to root aphid (*Pemphigus bursarius* L.) and downy mildew (*Bremia lactucae* Regel) in lettuce. *Euphytica* 29, 483–488.

Crute, I.R. and Harrison, J.M. (1988) Studies on the inheritance of resistance to metalaxyl in *Bremia lactucae* and on the stability and fitness of field isolates. *Plant Pathology* 37, 231–250.

Crute, I.R. and Johnson, A.G. (1976) The genetic relationship between races of *Bremia lactucae* and cultivars of *Lactuca sativa*. *Annals of Applied Biology* 83, 125–137.

Crute, I.R., and Norwood, J.M. (1981) The identification and characteristics of field resistance to lettuce downy mildew (*Bremia lactucae* Regel). *Euphytica* 30, 707–717.

Crute, I.R., Norwood, J.M. and Gordon, P.L. (1987) The occurrence, characteristics and distribution in the United Kingdom of resistance to phenylamide fungicides in *Bremia lactucae* (lettuce downy mildew). *Plant Pathology* 36, 297–315.

Currah, I.E., Gray, D. and Thomas, T.H. (1974) The sowing of germinating vegetable seeds using a fluid drill. *Annals of Applied Biology* 76, 311–318.

Curtis, I.S., He, C.P., Scott, R., Power, J.B. and Davey, M.R. (1996) Genomic male sterility in lettuce, a baseline for the production of F_1 hybrids. *Plant Science* 113, 113–119.

Daines, R.J., Minocha, S.C. and Kudakasseril, G.J. (1983) Induction of phenylalanine ammonia-lyase (PAL) in germinating lettuce seeds (*Lactuca sativa*). *Physiologia Plantarum* 59, 134–140.

Datnoff, L.E. and Nagata, R.T. (1992) Relationship between corky root disease and yield of iceberg lettuce. *Journal of the American Society for Horticultural Science* 117, 54–58.

Davis, R.M., Subbarao, K.V., Raid, R.N. and Kurtz, E.A. (eds) (1997) *Compendium of Lettuce Diseases*. APS Press, St Paul.

De Baynast, R. and Renard, C. (1994). Procédés d'extraction et de transformation des sucres de chicorée. *Comptes Rendus de l'Académie Agricole Française* 80, 31–46.

Demeulemeester, M.A.C., Rademacher, W., Van de Mierop, A. and De Proft, M.P. (1995) Influence of gibberelin synthesis inhibitors on stem elongation and floral initiation on *in vitro* chicory root explants under dark and light conditions. *Plant Growth Regulation* 17, 47–52.

Den Outer, R.W. (1989) Internal browning of witloof chicory (*Cichorium intybus* L.). *Journal of Horticultural Science* 64, 697–704.

Deslandes, J.A. (1954) Studies and observations on lettuce powdery mildew. *Plant Disease Reporter* 38, 560–562.

Desprez, B.F., Delesalle, L., Dhellemmes, C. and Desprez, M.F. (1994) Génétique et amélioration de la chicorée industrielle. *Comptes Rendus Académie Agricole Française* 80, 47–62.

DeVries, I.M. and van Raamsdonk, L.W.D. (1994) Numerical morphological analysis of lettuce cultivars and species (*Lactuca* setc. *Lactuca, Asteraceae*). *Plant Systematics and Evolution* 193, 125–141.

Diaz, C. and Hotchkiss, J.H. (1996) Comparative growth of *Escherichia coli* 0157:H7, spoilage organisms and shelf-life of shredded iceberg lettuce stored under modified atmospheres. *Journal of the Science of Food and Agriculture* 70, 433–438.

Dickson, M.H. (1963) Resistance to corky root rot in head lettuce. *Proceedings of the American Society for Horticultural Science* 82, 388–390.

Dickson, R.C. and Laird, E.F. Jr (1959) California desert and coastal populations of flying aphids and the spread of lettuce-mosaic virus. *Journal of Economic Entomology* 52, 440–443.

Dinant, S. and Lot, H. (1992) Lettuce mosaic virus: a review. *Plant Pathology* 41, 528–542.

Doerschug, M.R. and Miller, C.O. (1967) Chemical control of organ formation in *Lactuca sativa* explants. *American Journal of Botany* 54, 410–413.

Doré, C., Prigent, J. and Desprez, B. (1996) *In situ* gynogenetic haploid plants of chicory (*Cichorium intybus* L.) after intergeneric hybridization with *Cicerbita alpina* Walbr. *Plant Cell Reports* 15, 758–761.

Drake, V.C. (1948) Decline in viability of lettuce seed during laboratory storage. *Association of Official Seed Analysts News Letter* 22, 31–33.

Duffus, J.E. (1960) Radish yellows, a disease of radish, sugar beet, and other crops. *Phytopathology* 50, 389–394.

Duffus, J.E. (1961) Economic significance of beet western yellows (radish yellows) on sugar beet. *Phytopathology* 51, 605–607.

Duffus, J.E. (1963) Possible multiplication in the aphid vector of sowthistle yellow vein virus, a virus with an extremely long insect latent period. *Virology* 21, 194–202.

Duffus, J.E. (1965) Beet pseudo-yellow virus, transmitted by the greenhouse whitefly (*Trialeurodes vaporariorum*). *Phytopathology* 55, 450–453.

Duffus, J.E. (1972) Beet yellow stunt, a potentially destructive virus disease of sugar beet and lettuce. *Phytopathology* 62, 161–165.

Duffus, J.E. and Russell, G.E. (1972) Serological relationship between beet western yellows and turnip yellows viruses. *Phytopathology* 62, 1274–1277.

Duffus, J.E., Zink, F.W. and Bardin, R. (1970) Natural occurrence of sowthistle yellow vein virus on lettuce. *Phytopathology* 60, 1383–1384.

Duffus, J.E., Larsen, R.C. and Liu, H.Y. (1986) Lettuce infectious yellows virus – a new type of whitefly-transmitted virus. *Phytopathology* 76, 97–100.

Duffus, J.E., Liu, H.Y., Wisler, G.C. and Li, R.H. (1996) Lettuce chlorosis virus – a new whitefly-transmitted closterovirus. *European Journal of Plant Pathology* 102, 591–596.

Dujardin, M., Louant, B.P. and Tilquin, J.P. (1979) Determination du caryogramme de *Cichorium intybus* L. *Annales d'Amélioration des Plantes* 29, 305–310.

Dunn, J.A. (1959) The biology of lettuce root aphid. *Annals of Applied Biology* 47, 475–491.

Dunn, J.A. (1974) Study on inheritance of resistance to root aphid *Pemphigus bursarius*, in lettuce. *Annals of Applied Biology* 76, 9–18.

Durst, C.E. (1930) *Inheritance in Lettuce*. Bulletin 356. Illinois Agricultural Experiment Station, Urbana, Illinois, USA, 341 pp.

Dutta, S., Bradford, K.J. and Nevins, D.J. (1997) Endo-β-mannanase activity present in cell wall extracts of lettuce endosperm prior to radicle emergence. *Plant Physiology* 113, 155–161.

Eastwood, J. and Gray, D. (1976) New systems of establishing field lettuce crops assessed. *Horticultural Research* 15, 65–75.

Eenink, A.H. (1980) Breeding research on witloof chicory for the production of inbred lines and hybrids. In *Proceedings Eucarpia Meeting on Leafy Vegetables*. Glasshouse Crop Research Institution, Littlehampton, UK, pp. 5–11.

Eenink, A.H. (1981a) Compatibility and incompatibility in witloof chicory (*Cichorium intybus* L.). 1. The influence of temperature and plant age on pollen germination and seed production. *Euphytica* 30, 71–76.

Eenink, A.H. (1981b) Compatibility and incompatibility in witloof chicory (*Cichorium intybus* L.). 2. The incompatibility system. *Euphytica* 30, 77–85.

Eenink, A.H. (1982) Compatibility and incompatibility in witloof chicory (*Cichorium intybus* L.). 3. Gametic competition after mixed pollinations and double pollinations. *Euphytica* 31, 773–786.

Eenink, A.H. (1984) Compatibility and incompatibility in witloof chicory (*Cichorium intybus* L.). 4. Formation of self-seeds on a self-incompatible and a moderately self-compatible genotype after double and triple pollinations. *Euphytica* 33, 161–167.

Eenink, A.H. and Dieleman, F.L. (1977) Screening *Lactuca* for resistance to *Myzus persicae*. *Netherlands Journal of Plant Pathology* 83, 139–151.

Eenink, A.H. and Dieleman, F.L. (1980) Development of *Myzus persicae* on a partially resistant and on a susceptible genotype of lettuce (*Lactuca sativa*) in relation to plant age. *Netherlands Journal of Plant Pathology* 86, 111–116.

Eenink, A.H. and Dieleman, F.L. (1983) Inheritance of resistance to the leaf aphid *Nazonovia ribis-nigri* in the wild lettuce species *Lactuca virosa*. *Euphytica* 32, 691–695.

Eenink, A.H. and Garretsen, F. (1980) Research on the inheritance of fasciation in lettuce (*Lactuca sativa*). *Euphytica* 29, 653–660.

Eenink, A.H., Groenwold, R. and Dieleman, F.L. (1982a) Resistance of lettuce (*Lactuca*) to the leaf aphid *Nasonovia ribisnigri*. 1. Transfer of resistance from *L. virosa* to *L. sativa* by interspecific crosses and selection of resistant breeding lines. *Euphytica* 31, 291–300.

Eenink, A.H., Groenwold, R. and Dieleman, F.L. (1982b) Resistance of lettuce (*Lactuca*) to the leaf aphid *Nasonovia ribisnigri*. 2. Inheritance of the resistance. *Euphytica* 31, 301–304.

Elia, M. and Piglionica, V. (1964) Osservazioni preliminari sulla resistenza di cultivars di lattuga ai 'marciumi del colletto' da *Sclerotinia* spp. *Phytopathologia Mediterranea* 3, 37–39.

Ellis, P.R., Pink, D.A.C. and Ramsey, A.D. (1994) Inheritance of resistance to lettuce root aphid in the lettuce cultivars 'Avoncrisp' and 'Lakeland'. *Annals of Applied Biology* 124, 141–151.

Ellis, R.H., Hong, T.D. and Roberts, E.H. (1989) Response of seed germination in three genera of Compositae to white light of varying photon flux density and photoperiod. *Journal of Experimental Botany* 40, 13–22.

Engler, D.E. and Grogan, R.G. (1984) Variation in lettuce plants regenerated from protoplasts. *Journal of Heredity* 75, 426–430.

Ernst, M., Chatterton, N.J. and Harrison, P.A. (1995) Carbohydrate changes in chicory (*Cichorium intybus* L. var. *foliosum*) during growth and storage. *Sciencia Horticulturae* 63, 251–261.

Ernst-Schwarzenbach, M. (1932) Zur Genetik und Fertilitat von *Lactuca sativa* L. und *Cichorium endivia* L. *Archiv der Julius Klaus-Stiftung* 7, 1–35.

Errampalli, D., Fletcher, J. and Claypool, P.L. (1991) Incidence of yellows in carrot and lettuce and characterization of mycoplasmalike organism isolates in Oklahoma. *Plant Disease* 75, 579–584.

Eurostat (1998) *Crop Production Statistics*. Data Shop Eurostat, Brussels, Belgium.

Falk, B.W. and Guzman, V.L. (1981) A virus as the causal agent of spring yellows of lettuce and escarole. *Proceedings of the Florida State Horticultural Society* 94, 149–152.

Falk, B.W. and Purcifull, D.E. (1983) Development and application of an enzyme-linked immunosorbent assay (ELISA) test to index lettuce seeds for lettuce mosaic in Florida. *Plant Disease* 67, 413–416.

Falk, B.W., Duffus, J.E. and Morris, T.J. (1979) Transmission, host range, and serological properties of the viruses that cause lettuce speckles disease. *Phytopathology* 69, 612–617.

Falk, B.W., Purcifull, D.E. and Christie, S.R. (1986) Natural occurrence of sonchus yellow net virus. *Plant Disease* 70, 591–593.

Fanizza, G. and Damato, G. (1995) Analysis of seed yield components in chicory (*Cichorium intybus* L.). *Journal of Applied Seed Production* 13, 22–24.

Farrara, B.F. and Michelmore, R.W. (1987) Identification of new sources of resistance to downy mildew in *Lactuca* spp. *HortScience* 22, 647–649.

Farrara, B.F., Ilott, T.W. and Michelmore, R.W. (1987) Genetic analysis of factors for resistance to downy mildew (*Bremia lactucae*) in species of lettuce (*Lactuca sativa* and *L. serriola*). *Plant Pathology* 36, 499–514.

Feráková, V. (1977) *The Genus* Lactuca *L. in Europe.* Universita Komenskeho, Bratislava.

Fielding, A., Kristie, D.N. and Dearman, P. (1992) The temperature dependence of Pfr action governs the upper limit for germination in lettuce. *Photochemistry and Photobiology* 56, 623–627.

Flint, L.H. and McAlister, E.D. (1937) Wave lengths of radiation in the visible spectrum promoting the germination of light-sensitive lettuce seed. *Smithsonian Miscellaneous Collections* 96, 1–8.

Forbes, A.R. and Mackenzie, J.R. (1982) The lettuce aphid, *Nasonovia ribisnigri* (Homoptera: Aphididae) damaging lettuce crops in British Columbia. *Journal of the Entomological Society of British Columbia* 79, 28–31.

Fouldrin, K. and Limami, A. (1993) The influence of nitrogen ($^{15}NO_3$) supply to chicory (*Cichorium intybus* L.) plants during forcing on the uptake and remobilization of N reserves for chicon growth. *Journal of Experimental Botany* 44, 1313–1319.

Frese, L. and Dambroth, M. (1987) Research on the genetic resources of inulin-containing chicory (*Cichorium intybus*). *Plant Breeding* 99, 308–317.

Fry, P.R., Close, R.C., Procter, C.H. and Sunde, R. (1973) Lettuce necrotic yellows virus in New Zealand. *New Zealand Journal of Agricultural Research* 16, 143–146.

Gallardo, M., Jackson, L.E., Schulbach, K., Snyder, R.L., Thompson, R.B. and Wyland, L.J. (1996a) Production and water use in lettuce under variable water supply. *Irrigation Science* 16, 125–137.

Gallardo, M., Jackson, L.E. and Thompson, R.B. (1996b) Shoot and root physiological responses to localized zones of soil moisture in cultivated and wild lettuce (*Lactuca* spp.) *Plant, Cell, and Environment* 19, 1169–1178.

Galli, M.G. (1988) The role of DNA synthesis during hypocotyl elongation in light and dark. *Annals of Botany* 62, 287–293.

Gallitelli, D. and Di Franco, A. (1982) Chicory virus X: a newly recognized potexvirus of *Cichorium intybus*. *Phytopathologie Zeitschrift* 105, 120–130.

Gardner, G. (1983) The effect of growth retardants on phytochrome-induced lettuce seed germination. *Journal of Plant Growth Regulation* 2, 159–163.

Garibaldi, A. and Tesi, R. (1971) Resistance to *Alternaria porri* f. sp. *cichorii* in *Cichorium* species and its heredity. *Revista di Ortoflorofrutticoltura* 55, 350–355.

Gates, R.R. and Rees, E.M. (1921) A cytological study of pollen development in *Lactuca*. *Annals of Botany (London)* 35, 365–398.

Gaudreau, L., Charbonneau, J., Vezina, L.P. and Gosselin, A. (1995) Effects of photoperiod and photosynthetic photon flux on nitrate content and nitrate reductase activity in greenhouse-grown lettuce. *Journal of Plant Nutrition* 18, 437–453.

George, R.A.T. (1985) *Vegetable Seed Production*. Longman, New York.

Gerstel, D.U. (1950) Self-incompatibility studies in guayule II. Inheritance. *Genetics* 35, 482–506.

Gianquinto, G. and Pimpini, F. (1989) The influence of temperature on growth, bolting and yield of chicory cv. Rosso di Chioggia (*Cichorium intybus* L.). *Journal of Horticultural Science* 64, 687–695.

Gianquinto, G. and Pimpini, F. (1995) Morphological and physiological aspects of phase transition in radicchio (*Cichorium intybus* L. var. *sylvestre* Bischoff): the influence of temperature. *Advances in Horticultural Science* 9, 192–199.

Glenn, E.P. (1984) Seasonal effects of radiation and temperature on growth of greenhouse lettuce in a high insolation desert environment. *Scientia Horticulturae* 22, 9–21.

Globerson, D. (1981) The quality of lettuce seed harvested at different times after anthesis. *Seed Science and Technology* 9, 861–866.

Globerson, D. and Ventura, J. (1973) Influence of gibberellins on promoting flowering and seed yield in bolting-resistant lettuce cultivars. *Israel Journal of Agricultural Research* 23, 75–77.

González, L. and López Ch., R. (1980) Efecto de densidades de inoculo y caracteristicas del suelo sobre la patogenicidad de *Meloidogyne incognita* en lechuga. *Agronomia Costarricense* 4, 155–163.

Gorsk, T. and Gorska, K. (1979) Inhibitory effects of full daylight on the germination of *Lactuca sativa* L. *Planta* 144, 121–124.

Gray, D. (1975) Effects of temperature on the germination and emergence of lettuce (*Lactuca sativa* L.) varieties. *Journal of Horticultural Science* 50, 349–361.

Gray, D., Steckel, J.R.A., Wurr, D.C.E. and Fellows, J.R. (1986) The effects of applications of gibberellins to the parent plant, harvest date and harvest method on seed yield and mean seed weight of crisphead lettuce. *Annals of Applied Biology* 108, 125–134.

Gray, D., Wurr, D.C.E., Ward, J.A. and Fellows, J.R. (1988) Influence of post-flowering temperature on seed development and subsequent performance of crisp lettuce. *Annals of Applied Biology* 113, 391–402.

Grogan, R.G. (1980) Control of lettuce mosaic with virus-free seed. *Plant Disease* 64, 446–449.

Grogan, R.G. and Schnathorst, W.C. (1955) Tobacco ring-spot virus – the cause of lettuce calico. *Plant Disease Reporter* 39, 803–806.

Grogan, R.G., Welch, J.E. and Bardin, R. (1952) Common lettuce mosaic and its control by the use of mosaic-free seed. *Phytopathology* 42, 573–578.

Grogan, R.G., Snyder, W.C. and Bardin, R. (1955) *Diseases of Lettuce*. Circular 448, California Agricultural Experiment Station, Berkeley, California, USA, 28 pp.

Grogan, R.G., Zink, F.W., Hewitt, W.B. and Kimble, K.A. (1958) The association of *Olpidium* with the big-vein disease of lettuce. *Phytopathology* 48, 292–296.

Grogan, R.G., Misaghi, I.J., Kimble, K.A., Greathead, A.S., Ririe, D. and Bardin, R. (1977) Varnish spot, destructive disease of lettuce in California caused by *Pseudomonas cichorii. Phytopathogy* 67, 957–960.

Gustafsson, I. (1992) Race non-specific resistance to *Bremia lactucae. Annals of Applied Biology* 120, 127–136.

Haber, A.H. and Tolbert, N.E. (1959) Effects of gibberellic acid, kinetin and light on the germination of lettuce seed. In: Withrow, R.W. (ed.) *Photoperiodism and Related Phenomena in Plants and Animals*. American Association for the Advancement of Science, Washington, DC., pp. 197–205.

Hagenmaier, R.D. and Baker, R.A. (1997) Low-dose irradiation of cut iceberg lettuce in modified atmosphere packaging. *Journal of Agricultural Food Chemistry* 45, 2864–2868.

Hand, D.W., Sweeney, D.G., Hunt, R. and Wilson, J.W. (1985) Integrated analysis of growth and light interception in winter lettuce II. Differences between cultivars. *Annals of Botany* 56, 673–682.

Harlan, J. (1986) Lettuce and the sycomore: sex and romance in Ancient Egypt. *Economic Botany* 40, 4–15.

Harrington, J.F. (1960) The use of gibberellic acid to induce bolting and increase seed yield of tight-heading lettuce. *Proceedings of the American Society for Horticultural Science* 75, 476–479.

Harrington, J.F., Rappaport, L. and Hood, K.J. (1957) Influence of gibberellins on stem elongation and flowering of endive. *Science* 125, 601–602.

Harris, M.R. (1939) A survey of spotted wilt disease of lettuce in the Salinas Valley. *Bulletin of the Department of Agriculture, State of California* 28, 201–213.

Harsh, D.D., Vyas, O.P., Bohra, S.P. and Sankhlu, N. (1973) Lettuce seed germination: prevention of thermodormancy by 2-chloroethanephosphonic acid (ethane). *Experientia* 731–732.

Hawthorn, L.R. and Pollard, L.H. (1951) Selection for Great Lakes lettuce strains for higher seed yields. *Proceedings of the American Society for Horticultural Science* 57, 323–328.

Hawthorn, L.R. and Pollard, L.H. (1956) *Production of Lettuce Seed as affected by Soil Moisture and Fertility*. Bulletin 386, Utah State Agricultural College, Logan, Utah, USA, pp. 2–23.

Hawthorne, B.T. (1975) Effect of mulching on the incidence of *Sclerotinia minor* in lettuce. *New Zealand Journal of Experimental Agriculture* 3, 273–274.

Hedrick, U.P. (ed.) (1972) *Sturtevant's Edible Plants of the World.* Dover, New York.

Heimdal, H., Kühn, B.F., Poll, L. and Larsen, L.M. (1995) Biochemical changes and sensory quality of shredded and MA-packaged iceberg lettuce. *Journal of Food Science* 60, 1265–1278.

Heydecker, W. and Joshua, A. (1976) Delayed interacting effects of temperature and light on the germination of *Lactuca sativa* seeds. *Seed Science Technology* 4, 231–238.

Hicks, J.R. and Hall, C.B. (1972) Control of shredded lettuce discoloration. *Florida State Horticultural Society* 85, 219–221.

Hikichi, Y., Saito, A. and Suzuki, K. (1996) Infection sites of *Pseudomonas cichorii* into head leaf of lettuce. *Annals of the Phytopathology Society of Japan* 62, 125–129.

Huang, X.L. and Khan, A.A. (1988) Post-germination root growth in lettuce: role of ethylene under saline and nonsaline conditions. *HortScience* 23, 1040–1042.

Hubbard, J.C. and Gerik, J.S. (1993) A new wilt disease of lettuce incited by *Fusarium oxysporum* f. sp. *lactucum* forma specialis nov. *Plant Disease* 77, 750–754.

Hughes, M.B. and Babcock, E.B. (1950) Self-incompatibility in *Crepis foetida* (L.) Subsp. *Rhoeadifolia* Schinz et Keller. *Genetics* 35, 570–588.

Hulbert, S.H. and Michelmore, R.W. (1985) Linkage analysis of genes for resistance to downy mildew (*Bremia lactucae*) in lettuce (*Lactctuca sativa*). *Theoretical and Applied Genetics* 70, 520–528.

Hunt, R., Wilson, J.W., Hand, D.W. and Sweeney, D.G. (1984) Integrated analysis of growth and light interception in winter lettuce I. Analytical methods and environmental influences. *Annals of Botany* 54, 743–757.

Hutin, C. (1995) Production et échanges européens de laitues. *Infos* 114, 14–18.

Huyskes, J.A. (1962) Cold requirements of witloof chicory varieties (*Cichorium intybus* L.) as a yield determining factor. *Euphytica* 11, 36–41.

Huyskes, J.A. (1963) Veredeling van witloof voor het trekken zonder dekgrond. *Instituut voor de Veredeling van Tuinbouwgewassen* 202, 1–70.

Ibrahim, A.E., Roberts, E.H. and Murdoch, A.J. (1983) Viability of lettuce seeds. II. Survival and oxygen uptake in osmotically controlled storage. *Journal of Experimental Botany* 34, 631–640.

Ikuma, H. and Thimann, K.V. (1964) Analysis of germination processes of lettuce seed by means of temperature and anaerobiosis. *Plant Physiology* 39, 756–767.

Imolehin, E.D. and Grogan, R.G. (1980) Factors affecting survival of sclerotia, and effects of inoculum density, relative position, and distance of sclerotia from the host on infection of lettuce by *Sclerotinia minor. Phytopathology* 70, 1162–1167.

Imolehin, E.D., Grogan, R.G. and Duniway, J.M. (1980) Effect of temperature and moisture tension on growth, sclerotial production, germination, and infection by *Sclerotinia minor. Phytopathology* 70, 1153–1157.

Inada, K. and Yabumoto, Y. (1989) Effects of light quality, daylength and periodic temperature variation on the growth of lettuce and radish plants. *Japanese Journal of Crop Science* 58, 689–694.

Inoue, Y. and Nagashima, H. (1991) Photoperceptive site in phytochrome-mediated lettuce (*Lactuca sativa* L. cv. Grand Rapids) seed germination. *Journal of Plant Physiology* 137, 669–673.

Isenberg, F.M. and Hartman, J. (1958) *Vacuum Cooling Vegetables.* Cornell Extension

Bulletin 1012, New York State College of Agriculture, Ithaca, New York, USA, pp. 2–19.

Izzeldin, H., Lippert, L.F. and Takatori, F.H. (1980) An influence of water stress at different growth stages on yield and quality of lettuce seed. *Journal of the American Society for Horticultural Science* 105, 68–71.

Jackson, L.E. (1995) Root architecture in cultivated and wild lettuce (*Lactuca* spp.). *Plant, Cell, and Environment* 18, 885–897.

Jacques, M.A., Kinkel, L.L. and Morris, C.E. (1995) Population sizes, immigration, and growth of epiphytic bacteria on leaves of different ages and positions of field-grown endive (*Cichorium endivia* var. *latifolia*). *Applied and Environmental Microbiology* 61, 899–906.

Jagger, I.C. (1921) A transmissible mosaic disease of lettuce. *Journal of Agricultural Research* 20, 737–741.

Jagger, I.C. and Chandler, N. (1934) Big vein of lettuce. *Phytopathology* 24, 1253–1256.

Jagger, I.C. and Whitaker, T.W. (1940). The inheritance of immunity from mildew (*Bremia lactucae*) in lettuce. *Phytopathology* 30, 427–433.

Jagger, I.C., Whitaker, T.W., Uselman, J.J. and Owen, W.M. (1941) *The Imperial Strains of Lettuce*. No. 596, United States Department of Agriculture, Washington, DC USA, 15 pp.

Jenkins, J.M., Jr (1962) Brown rib resistance in lettuce. *Proceedings of the American Society for Horticultural Science* 81, 376–378.

Johnson, A.G., Crute, I.R. and Gordon, P.L. (1977) The genetics of race specific resistance in lettuce (*Lactuca sativa*) to downy mildew (*Bremia lactucae*). *Annals of Applied Biology* 86, 87–103.

Johnson, A.G., Laxton, S.A., Crute, I.R., Gordon, P.L. and Norwood, J.M. (1978) Further work on the genetics of race specific resistance in lettuce (*Lactuca sativa*) to downy mildew (*Bremia lactucae*). *Annals of Applied Biology* 89, 257–264.

Johnson, M.W., Kido, K., Toscano, N.C., Van Steenwyk, R.A., Welter, S.C. and McCalley, N.F. (1984) Strategies for managing lepidopterous pests on lettuce. *California Agriculture* 38 (1 and 2), 6–8.

Jones, H.A. (1927) Pollination and life history studies of lettuce (*Lactuca sativa* L.). *Hilgardia* 2, 425–479.

Joseph, C. (1986) The cytokinins of *Cichorium intybus* L. root: identification and changes during vernalization. *Journal of Plant Physiology* 124, 235–246.

Joseph, C., Seigneuret, J.M., Touraud, G. and Billot, J. (1983) The free gibberellins of *Cichorium intybus* L. root: identification and changes during vernalization. *Zeitschrift fur Pflanzenphysiologie* 110, 401–407.

Junttila, O. and Stushnoff, C. (1977) Freezing avoidance by deep supercooling in hydrated lettuce seeds. *Nature* 269, 325–327.

Kader, A.A., Brecht, P.E., Woodruff, R. and Morris, L.L. (1973) Influence of carbon monoxide, carbon dixoide, and oxygen levels on brown stain, respiration rate, and visual quality of lettuce. *Journal of the American Society for Horticultural Science* 98, 485–488.

Kahn, A., Goss, J.A. and Smith, D.E. (1957) Effect of gibberellin on germination of lettuce seed. *Science* 125, 645–646.

Kassanis, B. (1947) Studies on dandelion yellow mosaic and other virus diseases of lettuce. *Annals of Applied Biology* 34, 412–421.

Ke, D.Y. and Saltveit, M.E., Jr. (1986) Effects of calcium and auxin on russet spotting and phenylalanine ammonia-lyase activity in iceberg lettuce. *HortScience* 21, 1169–1171.

Keimer, L. (1924) *Die Gartenpflanzen in Alten Aegypten.* Hoffmann und Campe, Hamburg.

Kesseli, R.V., Ochoa, O. and Michelmore R. (1991) Variation in RFLP loci in *Lactuca* spp. and origin of cultivated lettuce (*L. sativa*). *Genome* 34, 430–436.

Kesseli, R.V., Paran, I. and Michelmore, R.W. (1994) Analysis of a detailed genetic linkage map of *Lactuca sativa* (lettuce) constructed from RFLP and RAPD markers. *Genetics* 136, 1435–1446.

King, A.D., Jr, Magnuson, J.A. and Török, T. (1991) Microbial flora and storage quality of partially processed lettuce. *Journal of Food Science* 56, 459–461.

Kishaba, A.N., McCreight, J.D., Coudriet, D.J., Whitaker, T.W. and Pesho, G.R. (1980) Studies of ovipositional preference of cabbage looper on progenies form a cross between cultivated and prickly lettuce. *Journal of the American Society for Horticultural Science* 105, 890–892.

Klaasen, V.A., Boeshore, M.L., Koonin, E.V., Tian, T.Y. and Falk, B.W. (1995) Genome structure and phylogenetic analysis of lettuce infectious yellows virus, a whitefly-transmitted bipartite closterovirus. *Virology* 208, 99–110.

Knight, S.L. and Mitchell, C.A. (1988) Growth and yield characteristics of 'Waldmann's Green' leaf lettuce under different photon fluxes from metal halide or incandescent + fluorescent radiation. *Scientia Horticulturae* 35, 51–61.

Kordan, H.A. (1981) Concentration–time related effects of colchicine on lettuce hypocotyl growth. *Zeitschrift Pflanzenphysiologie* 102, 379–388.

Kosar, W.F. and Thompson, R.C. (1957) Influence of storage humidity on dormancy and longevity of lettuce seed. *Proceedings of the American Society for Horticultural Science* 70, 273–276.

Krahnstover, K., van Kruistum, G., Lips, J. and Sarrazyn, R. (1997) Physiologische Anomalien bei Chicoree. *Gemuse (Munchen)* 33, 506–509.

Kristie, D.N., Bassi, P.K. and Spencer, M.S. (1981) Factors affecting the induction of secondary dormancy in lettuce. *Plant Physiology* 67, 1224–1229.

Kruger, N.S. (1966) Tipburn of lettuce in relation to calcium nutrition. *Queensland Journal of Agricultural and Animal Science* 23, 379–385.

Kunkel, L.O. (1926) Studies on aster yellows. *American Journal of Botany* 13, 646–705.

Künsch, U., Schärer, H. and Hurter, J. (1995) Qualitätsuntersuchungen an Kopf- und Nüsslisalat aus Horssol- und konvetionellem Glashausanbau. *Mitteilung Gebiete Lebensmitteluntersuchung Hygienisch* 86, 637–647.

Kuwata, S., Kubo, S., Yamashita, S. and Doi, Y. (1983) Rod-shaped particles, a probable entity of lettuce big vein virus. *Annals of the Phytopathology Society of Japan* 49, 246–251.

Lavigne, C., Millecamps, J.L., Manach, H., Cordonnier, P., Matejicek, A., Vasseur, J. and Gasquez, J. (1994) Monogenetic semidominant sulfonylurea resistance in a line of white chicory. *Plant Breeding* 113, 305–311.

Lavigne, C., Manac'h, H., Guyard, C. and Gasquez, J. (1995) The cost of herbicide resistance in white-chicory: ecological implications for its commercial release. *Theoretical and Applied Genetics* 91, 1301–1308.

Lebeda, A. (1994) Evaluation of wild *Lactuca* species for resistance of natural infection of powdery mildew (*Erysiphe cichoracearum*). *Genetic Resources and Crop Evolution* 41, 55–57.

Lee, P.E. and Robinson, A.G. (1958) Studies on the six-spotted leafhopper, *Macrosteles fascifrons* (Stal.) and aster yellows in Manitoba. *Canadian Journal of Plant Science* 38, 320–327.

Lenker, D.H. and Adrian, P.A. (1971) Use of X-rays for selecting mature lettuce heads. *Transactions of the American Society of Agricultural Engineers* 14, 894–898.

Lenker, D.H., Adrian, P.A., French, G.W. and Zahara, M. (1973) Selective mechanical lettuce harvesting system. *Transactions of the American Society of Agricultural Engineers* 16, 858–861, 866.

Leroux, M. (1994) Les variétés de chicorée industrielle face aux exigences de qualité du marché. *Comptes Rendus Académie Agricole Française* 80, 69–82.

Lewis, M.T. (1931) Inheritance of heading characteristics in lettuce varieties. *Proceedings of the American Society for Horticultural Science* 27, 347–351.

Lindqvist, K. (1958) Inheritance of lobed leaf form in *Lactuca*. *Hereditas* 44, 347–377.

Lindqvist, K. (1960a) Cytogenetic studies in the *serriola* group of *Lactuca*. *Hereditas* 46, 75–151.

Lindqvist, K. (1960b) On the origin of cultivated lettuce. *Hereditas* 46, 319–350.

Lindqvist, K. (1960c) Inheritance studies in lettuce. *Hereditas* 46, 387–470.

Lips, J. (1976) Some experiences with new varieties of witloof for forcing without soil covering. In: *Proceedings Eucarpia Meeting on Leafy Vegetables*. Institute for Horticultural Plant Breeding, Wageningen, Netherlands, pp. 40–47.

López-Gálvez, G., Saltveit, M. and Cantwell, M. (1996) The visual quality of minimally processed lettuces stored in air or controlled atmosphere with emphasis on romaine and iceberg types. *Postharvest Biology and Technology* 8, 179–190.

Louant, B.P., Plumier, W., Dôme, J. and Aussems, A. (1978) Critères d'identification, de purification et de sélection chez *Cichorium intybus* L. (Chicorée de Bruxelles) 1. Observations préliminaires sur quelques caractéristiques de l'akène. *Bulletin Recherches Agronomique Gembloux* 13, 59–72.

McCreight, J.D. (1987) Resistance in wild lettuce to lettuce infectious yellows. *HortScience* 22, 640–642.

McCreight, J.D., Kishaba, A.N. and Mayberry, K.S. (1986) Lettuce infectious yellows tolerance in lettuce. *Journal of the American Society for Horticultural Science* 111, 788–792.

MacIsaac, S.A., Sawhney, V.K. and Pohorecky, Y. (1989) Regulation of lateral root formation in lettuce (*Lactuca sativa*) seedling roots: interacting effects of α-naphthaleneacetic acid and kinetin. *Physiologia Plantarum* 77, 287–293.

McKinney, K.B. (1944) *The Cabbage Looper as a Pest of Lettuce in the Southwest*. Technical Bulletin 846, United States Department of Agriculture, Washington, DC, USA, 30 pp.

McLean, D.L. (1962) Transmission of lettuce mosaic virus by a new vector, *Pemphigus bursarius*. *Journal of Economic Entomology* 55, 580–583.

Magnuson, J.A., King, A.D., Jr. and Török, T. (1990) Microflora of partially processed lettuce. *Applied and Environmental Microbiology* 56, 3851–3854.

Maisonneuve, B., Chovelon, V. and Lot, H. (1991) Inheritance of resistance to beet western yellows virus in *Lactuca virosa* L. *HortScience* 26, 1543–1545.

Maisonneuve, B., Chupeau, M.C., Bellec, Y. and Chupeau, Y. (1995) Sexual and somatic hybridization in the genius *Lactuca*. *Euphytica* 85, 281–285.

Mallory-Smith, C.A., Thill, D.C. and Dial, M.J. (1990) Identification of sulfonylurea herbicide-resistant prickly lettuce (*Lactuca serriola*). *Weed Technology* 4, 163–168.

Marcum, D.B., Grogan, R.G. and Greathead, A.S. (1977) Fungicide control of lettuce drop caused by *Sclerotinia sclerotiorum* 'minor'. *Plant Disease Reporter* 61, 555–559.

Marlatt, R.B. (1974) *Nonpathaogenic Diseases of Lettuce, Their Identification and Control.* Bulletin 721A, University of Florida Agricultural Experiment Stations, Gainsville, Florida, USA, 47 pp.

Marlatt, R.B. and Stewart, J.K. (1956) Pink rib of head lettuce. *Plant Disease Reporter* 40, 742–743.

Martin, F.W. and Ruberte, R.M. (1975) *Edible Leaves of the Tropics.* Agency for International Development, Department of State, and Agricultural Research Service, US Department of Agriculture, Mayaguez, Puerto Rico.

Mayhew, D.E. and Matsumoto, T.T. (1978) Romaine lettuce, a new host for tobacco rattle virus. *Plant Disease Reporter* 62, 553–556.

Maynard, D.N., Gersten, B. and Vernell, H.F. (1962) The cause and control of brownheart of escarole. *Proceedings of the American Society for Horticultural Science* 81, 371–375.

Michelmore, R.W. (1995) Isolation of disease resistance genes from crop plants. *Current Opinion in Biotechnology* 6, 145–152.

Michelmore, R.W. and Eash, J.A. (1986) Lettuce. In: Evans, D.A., Sharp, W.R. and Amirato, P.V. (eds) *Handbook of Plant Cell Culture,* Vol. 4. MacMillan, New York, pp. 512–551.

Michelmore, R.W. and Ingram, D.S. (1980) Heterothallism in *Bremia lactucae. Transactions of the British Mycological Society* 75, 47–56.

Michelmore, R.W. and Ingram, D.S. (1981) Recovery of progeny following sexual reproduction of *Bremia lactucae. Transactions of the British Mycological Society* 77, 131–137.

Michelmore, R.W., Marsh, E., Seely, S. and Landry, B. (1987) Transformation of lettuce mediated by *Agrobacterium tumefaciens. Plant Cell Reports* 6, 439–442.

Mirkov, T.E. and Dodds, J.A. (1985) Association of double-stranded ribonucleic acids with lettuce big vein disease. *Phytopathology* 75, 631–635.

Misaghi, I.J., Grogan, R.G. and Westerlund, F.V. (1981a) A laboratory method to evaluate lettuce cultivars for tipburn tolerance. *Plant Disease* 65, 342–344.

Misaghi, I.J., Matyac, C.A. and Grogan, R.G. (1981b) Soil and foliar application of calcium chloride and calcium nitrate to control tipburn of head lettuce. *Plant Disease* 65, 821–822.

Mitchell, C.A., Leakakos, T. and Ford, T.L. (1991) Modification of yield and chlorophyll content in leaf lettuce by HPS radiation and nitrogen treatments. *HortScience* 26, 1371–1374.

Moline, H.E. and Lipton, W.J. (1987) *Market Diseases of Beets, Chicory, Endive, Escarole, Globe Artichokes, Lettuce, Rhubarb, Spinach, and Sweetpotatoes.* Agriculture Handbook Number 155, United States Department of Agriculture, Agricultural Research Service, Washington, DC, USA, 86 pp.

Moline, H.E. and Pollack, F.G. (1976) Conidiogenesis of *Marssonina panattoniana* and its potential as a serious postharvest pathogen of lettuce. *Phytopathology* 66, 669–674.

Moloney, S.C. and Milne, G.D. (1993) Establishment and management of Grasslands Puna chicory used as a specialist, high quality forage herb. *Proceedings of the New Zealand Grassland Association* 55, 113–118.

Nagata, R.T. (1992) Clip-and-wash method of emasculation for lettuce. *HortScience* 27, 907–908.

Neergard, P. and Newhall, A.G. (1951) Notes on the physiology and pathogenicity of *Centrospora acerina* (Hartig) Newhall. *Phytopathology* 41, 1021–1033.

Negm, F.B., Smith, O.E. and Kumamoto, J. (1972) Interaction of carbon dioxide and ethylene in overcoming thermodormancy of lettuce seeds. *Plant Physiology* 49, 869–872.

Netzer, D., Globerson, D. and Sacks, J. (1976) *Lactuca saligna* L., a new source of resistance to downy mildew (*Bremia lactucae* Reg.). *HortScience* 11, 612–613.

Netzer, D., Globerson, D., Weintal, C. and Elyassi, R. (1985) Sources and inheritance of resistance to Stemphylium leaf spot of lettuce. *Euphytica* 34, 393–396.

Newhall, A.G. (1923) Seed transmission of lettuce mosaic. *Phytopathology* 13, 104–106.

Newton, H.C. and Sequeira, L. (1972) Possible sources of resistance in lettuce to *Sclerotinia sclerotiorum*. *Plant Disease Reporter* 56, 875–878.

Norwood, J.M. and Crute, I.R. (1985) A comparison of the susceptibility of lettuce cultivars to natural field and artificially induced laboratory infection with downy mildew (*Bremia lactucae*). *Zeitschrift Pflanzenzuchtung* 95, 63–73.

Norwood, J.M., Crute, I.R. and Lebeda, A. (1981) The location and characteristics of novel sources of resistance to *Bremia lactucae* Regel (downy mildew) in wild *Lactuca* L. species. *Euphytica* 30, 659–668.

Norwood, J.M., Crute, I.R., Johnson, A.G. and Gordon, P.L. (1983a) A demonstration of the inheritance of field resistance to lettuce downy mildew (*Bremia lactucae* Regel) in progeny derived from cv. Grand Rapids. *Euphytica* 32, 161–170.

Norwood, J.M., Michelmore, R.W., Crute, I.R. and Ingram, D.S. (1983b) The inheritance of specific virulence in *Bremia lactucae* (downy mildew) to match resistance factors 1, 2, 4, 6, and 11 in *Lactuca sativa* (lettuce). *Plant Pathology* 32, 177–186.

Norwood, J.M., Johnson, A.G., O'Brien, M. and Crute, I.R. (1985) The inheritance of field resistance to lettuce downy mildew (*Bremia lactucae*) in the cross 'Avon Crisp' × 'Iceberg'. *Zeitschrift Pflanzenzuchtung* 94, 259–262.

Nothmann, J. (1976a) Morphology of head formation of cos lettuce (*Lactuca sativa* L. cv. Romana). 1. The process of hearting. *Annal of Botany* 40, 1067–1072.

Nothmann, J. (1976b) Morphology of head formation of cos lettuce (*Lactuca sativa* L. cv. Romana). 2. The development of spiral-leaved heads. *Annals of Botany* 40, 1073–1077.

Nothmann, J. (1977a) Effects of soil temperature on head development of cos lettuce. *Scientia Horticulturae* 7, 97–105.

Nothmann, J. (1977b) Morphogenetic effects of seasonal conditions on head development of cos lettuce (*Lactuca sativa* L. var. *romana*). *Journal of Horticultural Science* 52, 155–162.

O'Brien, R.G. and Davis, R.D. (1994) Lettuce black root rot – a disease caused by *Chalara elegans*. *Australasian Plant Pathology* 23, 106–111.

O'Brien, R.D. and van Bruggen, A.H.C. (1992) Yield losses to iceberg lettuce due to corky root caused by *Rhizomonas suberifaciens*. *Phytopathology* 82, 154–159.

Ochoa, O., Delp, B. and Michelmore, R.W. (1987) Resistance in *Lactuca* spp. to *Microdochium panattoniana* (lettuce anthracnose). *Euphytica* 36, 609–614.

Oliver, G.W. (1910) *New Methods of Plant Breeding*. Bulletin 167, United States Bureau of Plant Industry, Washington, DC, USA, pp. 12–13.

Olivieri, A.M. (1972) Individuazione di alcuni parametri genetici in un incrocio di *Cichorium intybus* L. *Rivista di Agronomia* 6. 171–174.

O'Malley, P.J. and Hartmann, R.W. (1989) Resistance to tomato spotted wilt virus in lettuce. *HortScience* 24, 360–362.

Opgenorth, D.C., White, J.B., Oliver, B. and Greathead, A. (1991) Freeway daisy (*Osteospermum fruticosum*) as a host for lettuce mosaic virus. *Plant Disease* 75, 751.

Padhi, B. and Snyder, W.C. (1954) Stemphylium leaf spot of lettuce. *Phytopathology* 44, 175–180.

Paran, I. and Michelmore, R.W. (1993) Development of reliable PCR-based markers linked to downy mildew resistance genes in lettuce. *Theoretical and Applied Genetics* 85, 985–993.

Paran, I., Kesseli, R. and Michelmore, R. (1991) Identification of restriction fragment length polymorphism and random amplified polymorphic DNA markers linked to downy mildew resistance genes in lettuce, using near isogenic lines. *Genome* 34, 1021–1027.

Park, W.M. and Chen, S.S.C. (1974) Patterns of food utilization by the germinating lettuce seeds. *Plant Physiology* 53, 64–66.

Parman, Price, T.V. and Lee, M. (1991) Studies on fungicidal control of lettuce anthracnose. *Australasian Plant Pathology* 20, 103–107.

Pasternak, D., De Malach, Y., Borovic, I., Shram, M. and Aviram, C. (1986) Irrigation with brackish water under desert conditions IV. Salt tolerance studies with lettuce (*Lactuca sativa* L.). *Agricultural Water Management* II, 303–311.

Patterson, C.L. and Grogan, R.G. (1985) Differences in epidemiology and control of lettuce drop caused by *Sclerotinia minor* and *S. sclerotiorum*. *Plant Disease* 69, 766–770.

Patterson, C.L. and Grogan, R.G. (1991) Role of microsclerotia as primary inoculum of *Microdochium panattonianum*, incitant of lettuce anthracnose. *Plant Disease* 75, 134–138.

Pearson, O.H. (1956) The nature of the rogue in 456 lettuce. *Proceedings of the American Society for Horticultural Science* 68, 270–278.

Pearson, O.H. (1962) A simplified method for emasculating lettuce flowers. *Vegetable Improvement Newsletter* 4, 6.

Pécaut, P. (1962) Étude sur la système de reproduction de l'endive (*Cichorium intybus* L.). *Annales d'Amélioration des Plantes* 12, 265–296.

Perkins-Veazie, P. and Cantliffe, D.J. (1984) Need for high-quality seed for effective priming to overcome thermodormancy in lettuce. *Journal of the American Society for Horticultural Science* 109, 368–372.

Perring, T.M., Cooper, A.D., Rodriguez, R.J., Farrar, C.A. and Bellows, T.S., Jr (1993) Identification of a whitefly species by genomic and behavioral studies. *Science* 259, 74–77.

Pieczarka, D.J. and Lorbeer, J.W. (1974) Control of bottom rot of lettuce by ridging and fungicide application. *Plant Disease Reporter* 58, 837–840.

Pieczarka, D.J. and Lorbeer, J.W. (1975) Microorganisms associated with bottom rot of lettuce grown on organic soil in New York State. *Phytopathogy* 65, 16–21.

Pink, D.A.C., Walkey, D.G.A. and McClement, S.J. (1991) Genetics of resistance to beet western yellows virus in lettuce. *Plant Pathology* 40, 542–545.

Pink, D.A.C., Kostova, D. and Walkey, D.G.A. (1992a) Differentiation of pathotypes of lettuce mosaic virus. *Plant Pathology* 41, 5–12.

Pink, D.A.C., Lot, H. and Johnson, R. (1992b) Novel types of lettuce mosaic virus – breakdown of a durable resistance? *Euphytica* 63, 169–174.

Pollock, B.M. and Manalo, J.R. (1971) The influence of seed-lot history on sensitivity of lettuce seed to temperature and moisture stress. *HortScience* 6, 444–445.

Powers, N.J. (1995) Sticky short-run prices and vertical pricing: evidence from the market for iceberg lettuce. *Agribusiness* 11, 57–75.

Prabhaker, N., Coudriet, D.L. and Meyerdirk, D.E. (1985) Insecticide resistance in the sweetpotato whitefly, *Bemisia tabaci* (Homoptera: Aleyrodidae). *Journal of Economic Entomology* 78, 748–752.

Price, K.R., DuPont, M.S., Shepherd, R., Chan, H.W.S. and Fenwick, G.R. (1990) Relationship between the chemical and sensory properties of exotic salad crops – coloured lettuce (*Lactuca sativa*) and chicory (*Cichorium intybus*). *Journal of the Science of Food and Agriculture* 53, 185–192.

Prince, R.P. and Koontz, H.V. (1984) Lettuce production from a systems approach. In: *International Society of Soilless Culture Proceedings, Sixth International Congress*. Lunteren, the Netherlands, pp. 533–546.

Provvidenti, R., Robinson, R.W. and Shail, J.W. (1979) Chicory: a valuable source of resistance to turnip mosaic for endive and escarole. *Journal of the American Society for Horticultural Science* 104, 726–728.

Provvidenti, R., Robinson, R.W. and Shail, J.W. (1980) A source of resistance to a strain of cucumber mosaic virus in *Lactuca saligna* L. *HortScience* 15, 528–529.

Provvidenti, R., Robinson, R.W. and Shail, J.W. (1984) Incidence of broad bean wilt virus in lettuce in New York State and sources of resistance. *HortScience* 19, 569–570.

Pryor, D.E. (1941) A unique case of powdery mildew on lettuce in the field. *Plant Disease Reporter* 25, 74.

Purdy, L.H. (1979) *Sclerotinia sclerotiorum*: history, diseases and symptomatology, host range, geographic distribution, and impact. *Phytopathology* 69, 875–880.

Raleigh, S. (1978) *Lettuce – Prices, Costs, and Margins*. Economics Statistics, and Cooperatives Service 209, United States Department of Agriculture, Washington, DC, USA, pp. 31–35.

Ramadan, A.A.S. (1976) Characteristics of prickly lettuce seed oil in relation to methods of extraction. *Die Nahrung* 20, 579–583.

Rappaport, L. and Wittwer, S.H. (1956) Flowering in head lettuce as influenced by seed vernalization, temperature, and photoperiod. *Proceedings of the American Society for Horticultural Science* 67, 429–437.

Reinert, R.A., Tingey, D.T. and Carter, H.B. (1972) Ozone induced foliar injury in lettuce and radish cultivars. *Journal of the American Society for Horticultural Science* 97, 711–714.

Reinink, K. (1991) Genetics of nitrate content of lettuce, 1: analysis of generation means. *Euphytica* 554, 83–92.

Reinink, K. and Dieleman, F.L. (1989a) Comparison of sources of resistance to leaf aphids in lettuce (*Lactuca sativa* L.). *Euphytica* 40, 21–29.

Reinink, K. and Dieleman, F.L. (1989b) Resistance in lettuce to the leaf aphids *Macrosiphum euphorbiae* and *Uroleucon sonchi*. *Annals of Applied Biology* 115, 489–498.

Reinink, K., Dieleman, F.L. and Groenwold, R. (1988) Selection of lines of lettuce (*Lactuca sativa* L.) with a high level of partial resistance to *Myzus persicae* (Sulzer). *Euphytica* 37, 241–245.

Reinink, K., Dieleman, F.L., Jansen, J. and Montenarie, A.M. (1989) Interactions between plant and aphid genotypes in resistance of lettuce to *Myzus persicae* and *Macrosiphum euphorbiae. Euphytica* 43, 215–222.

Reinink, K., Dieleman, F.L. and Groenwold, R. (1995) Inheritance of partial resistance to the leaf aphids *Macrosiphum euphorbiae* and *Uroleucon sonchi* in lettuce. *Annals of Applied Biology* 127, 413–424.

Reynolds, T. and Thompson, P.A. (1973) Effects of kinetin, gibberellins, and (±) abscisic acid on the germination of lettuce (*Lactuca sativa*). *Physiologia Plantarum* 28, 516–522.

Rick, C.M. (1953) Hybridization between chicory and endive. *Proceedings of the American Society for Horticultural Science* 62, 459–466.

Robbins, M.A., Witsenboer, H., Michelmore, R.W., Laliberte, J.F. and Fortin, M.G. (1994) Genetic mapping of turnip mosaic resistance in *Lactuca sativa. Theoretical and Applied Genetics* 89, 583–589.

Roberts, H.A., Hewson, R.T. and Ricketts, M.E. (1977) Weed competition in drilled summer lettuce. *Horticultural Research* 17, 39–45.

Robinson, R.W., McCreight, J.D. and Ryder, E.J. (1983) The genes of lettuce and closely related species. *Plant Breeding Reviews* 1, 267–293.

Rodenburg, C.M. (ed.) (1960) *Varieties of Lettuce.* Plant Breeding Institute, Wageningen, Holland.

Roorda van Eysinga, J.P.N.L. and Smilde, K.W. (1981) *Nutritional Disorders in Glasshouse Tomatoes, Cucumbers and Lettuce.* Centre for Agricultural Publishing and Documentation, Wageningen, pp. 79–109.

Roos, E.E. and Stanwood, P.C. (1981) Effects of low temperature, cooling rate, and moisture content on seed germination of lettuce. *Journal of the American Society for Horticultural Science* 106, 30–34.

Rubatzky, V.E. and Yamaguchi, M. (1997) *World Vegetables, Principles, Production, and Nutritive Values,* 2nd edn. Chapman and Hall, New York.

Rutherford, P.P. (1977) Changes during prolonged cold storage in the reducing sugars in chicory roots and their effects on the chicons produced after forcing. *Journal of Horticultural Science* 52, 99–103.

Ryder, E.J. (1963) An epistatically controlled pollen sterile in lettuce (*Lactuca sativa* L.). *Proceedings of the American Society for Horticultural Science* 83, 585–589.

Ryder, E.J. (1967) A recessive male sterility gene in lettuce (*Lactuca sativa* L.). *Proceedings of the American Society for Horticultural Science* 91, 366–368.

Ryder, E.J. (1968) *Evaluation of Lettuce Varieties and Breeding Lines for Resistance to Common Lettuce Mosaic.* Technical Bulletin 1391, United States Department of Agriculture, Agricultural Research Service, Washington, DC, USA, 8 pp.

Ryder, E.J. (1970) Inheritance of resistance to common lettuce mosaic. *Journal of the American Society for Horticultural Science* 95, 378–379.

Ryder, E.J. (1971) Genetic studies in lettuce (*Lactuca sativa* L.). *Journal of the American Society for Horticultural Science* 96, 826–828.

Ryder, E.J. (1973) Seed transmission of lettuce mosaic virus in mosaic resistant lettuce. *Journal of the American Society for Hortcultural Science* 98, 610–614.

Ryder, E.J. (1979a) 'Salinas' lettuce. *HortScience* 14, 283–284.

Ryder, E.J. (1979b) 'Vanguard 75' lettuce. *HortScience* 14, 284–286.

Ryder, E.J. (1983) Inheritance, linkage, and gene interaction studies in lettuce. *Journal of the American Society for Horticultural Science* 108, 985–991.

Ryder, E.J. (1985) Use of early flowering genes to reduce generation time in backcrossing, with specific application to lettuce breeding. *Journal of the American Society for Horticultural Science* 110, 570–573.

Ryder, E.J. (1988) Early flowering time in lettuce as influenced by a second flowering time gene and seasonal variation. *Journal of the American Society for Horticultural Science* 113, 456–460.

Ryder, E.J. (1996) Inheritance of chlorophyll deficiency traits in lettuce. *Journal of Heredity* 87, 314–318.

Ryder, E.J. and Duffus, J.E. (1966) Effects of beet western yellows and lettuce mosaic viruses on lettuce seed production, flowering, and other characters in the greenhouse. *Phytopathology* 56, 842–844.

Ryder, E.J. and Johnson, A.S. (1974) Mist depollination of lettuce flowers. *HortScience* 9, 584.

Ryder, E.J. and Robinson, B.J. (1995) Big-vein resistance in lettuce: identifying, selecting, and testing resistant cultivars and breeding lines. *Journal of the American Society for Horticultural Science* 120, 741–746.

Sanada, M., Sakamoto, Y., Hayashi, M., Mashiko, T., Okamoto, A. and Ohnishi, N. (1993) Celery and lettuce. In: Redenbaugh, K. (ed.) *Synseeds: Application of Synthetic Seeds to Crop Improvement.* CRC Press, Boca Raton, pp. 305–327.

Santamaria, P. and Elia, A. (1997) Producing nitrate-free endive heads: effect of nitrogen form on growth, yield, and ion composition of endive. *Journal of the American Society for Horticultural Science* 122, 140–145.

Sargent, J.A., Tommerup, I.C. and Ingram, D.S. (1973) The penetration of a susceptible lettuce variety by the downy mildew fungus *Bremia lactucae* Regel. *Physiological Plant Pathology* 3, 231–239.

Savary, S. (1983) Épidémiologie de la cercosporiose de la laitue (*Lactuca sativa* L.) en République de Côte-d'Ivoire: étude de quelques étapes du cycle épidémiologique. *Agronomie* 3, 903–910.

Scaife, A., Cox, E.F. and Morris, G.E.L. (1987) The relationship between shoot weight, plant density, and time during the propagaton of four vegetable species. *Annals of Botany* 59, 325–334.

Scaife, M.A. (1973) The early relative growth rates of six lettuce cultivars as affected by temperature. *Annals of Applied Biology* 74, 119–128.

Scherm, H. and van Bruggen, A.H.C. (1994) Weather variables associated with infection of lettuce by downy mildew (*Bremia lactucae*) in coastal California. *Phytopathology* 84, 860–865.

Schettini, T.M., Legg, E.J. and Michelmore, R.W. (1991) Insensitivity to metalaxyl in California populations of *Bremia lactucae* and resistance of California cultivars to downy mildew. *Phytopathology* 81, 64–70.

Schnathorst, W.C. (1960) Effects of temperature and moisture stress on the lettuce powdery mildew fungus. *Phytopathology* 50, 304–308.

Schnathorst, W.C. (1962) Comparative ecology of downy and powdery mildews of lettuce. *Phytopathology* 52, 41–46.

Schnathorst, W.C. and Bardin, R. (1958) Susceptibility of lettuce varieties and hybrids to powdery mildew (*Erysiphe cichoracearum*). *Plant Disease Reporter* 42, 1273.

Schoofs, J. and De Langhe, E. (1988) Chicory (*Cichorium intybus* L.). In: Bajaj, Y.P.S. (ed.) *Biotechnology in Forestry and Agriculture*, Vol. 6. Springer-Verlag, Berlin, Germany, pp. 294–321.

Sequeira, L. (1970) Resistance to corky root rot in lettuce. *Plant Disease Reporter* 54, 754–758.

Severin, H.H.P. (1929) Yellow disease of celery, lettuce, and other plants, transmitted by *Cicadula sexnotata* (Fall.). *Hilgardia* 3, 543–583.

Sexton, R.J. and Zhang, M.X. (1995) Can retailers depress lettuce prices at farm level. *California Agriculture* 49(3), 14–18.

Sexton, R.J. and Zhang, M. (1996) A model of price determination for fresh produce with application to California iceberg lettuce. *American Journal of Agricultural Economics* 78, 924–934.

Shannon, M.C. and McCreight, J.D. (1984) Salt tolerance of lettuce introductions. *HortScience* 19, 673–675.

Shannon, M.C., McCreight, J.D. and Draper, J.H. (1983) Screening tests for salt tolerance in lettuce. *Journal of the American Society for Horticultural Science* 108, 225–230.

Shear, C.B. (1975) Calcium-related disorders of fruits and vegetables. *HortScience* 10, 361–365.

Shoeb, Z.E., Osman, F., El-Kirdassy, Z.H.M. and Eissa, M.H. (1969) Studies and evaluation of Egyptian lettuce seeds *Lactuca scariola* L. *Grasas y Aceites* 20, 125–128.

Shrefler, J.W., Stall, W.M. and Dusky, J.A. (1996) Spiny amaranth (*Amaranthus spinosus* L.), a serious competitor to crisphead lettuce (*Lactuca sativa* L.). *HortScience* 31, 347–348.

Singh, B., Yang, C.C., Salunkhe, D.K. and Rahman, A.R. (1972) Controlled atmosphere storage of lettuce. 1. Effects on quality and the respiration rate of lettuce heads. *Journal of Food Science* 37, 48–51.

Sinska, I. (1988) Stimulation of dark germination of light sensitive lettuce seeds by polyamines. *Acta Physiologiae Plantarum* 10, 11–16.

Smith, M.T. (1989) The ultrastructure of physiological necrosis in cotyledons of lettuce seeds (*Lactuca sativa* L.). *Seed Science and Technology* 17, 453–462.

Smith, O.E., Yen, W.W.L. and Lyons, J.M. (1968) The effects of kinetin in overcoming high-temperature dormancy of lettuce seed. *Proceedings of the American Society for Horticultural Science* 93, 444–453.

Smith, O.E., Welch, N.C. and Little, T.M. (1973a) Studies on seed quality: I. Effect of seed size and weight on vigor. *Journal of the American Society for Horticultural Science* 98, 529–533.

Smith, O.E., Welch, N.C. and McCoy, O.D. (1973b) Studies on lettuce seed quality: II. Relationship of seed vigor to emergence, seedling weight, and yield. *Journal of the American Society for Horticultural Science* 98, 552–556.

Snyder, W.C., Bardin, R. and Grogan, R.G. (1952) Powdery mildew of lettuce reappears in Salinas Valley. *Plant Disease Reporter* 36, 321–322.

Soffer, H. and Smith, O.E. (1974a) Studies on lettuce seed quality: III. Relationships between flowering pattern, seed yield, and seed quality. *Journal of the American Society for Horticultural Science* 99, 114–117.

Soffer, H. and Smith, O.E. (1974b) Studies on lettuce seed quality: IV. Individually measured embryo and seed characteristics in relation to continuous plant growth (vigor) under controlled conditions. *Journal of the American Society for Horticultural Science* 99, 270–275.

Soffer, H. and Smith, O.E. (1974c) Studies on lettuce seed quality: V. Nutritional effects. *Journal of the American Society for Horticultural Science* 99, 459–463.

Sonneveld, C. and Voogt, S. (1973) The effects of steam sterilisation with steam–air mixtures on the development of some glasshouse crops. *Plant and Soil* 38, 415–423.

Stanghellini, M.E. and Kronland, W.C. (1982) Root rot of chicory caused by *Phymatotrichum cryptogea*. *Plant Disease* 66, 262–263.

Stanghellini, M.E. and Kronland, W.C. (1986) Yield loss in hydroponically grown lettuce attributed to subclinical infection of feeder rootlets by *Pythium dissotocum*. *Plant Disease* 70, 1053–1056.

Stanghellini, M.E., Adaskaveg, J.E. and Rasmussen, S.L. (1990) Pathogenesis of *Plasmodiopara lactucae-radicis*, a systemic root pathogen of cultivated lettuce. *Plant Disease* 74, 173–178.

Steadman, J.R. (1979) Control of plant disease causd by *Sclerotinia* species. *Phytopathology* 69, 904–906.

Steiner, J.J. and Opoku-Boateng, K. (1991) Natural season-long and diurnal temperature effects on lettuce seed production and quality. *Journal of the American Society for Horticultural Science* 116, 396–400.

Stevens, M.A. (1974) Varietal influence on nutritional value. In: White, P.L. and Selvey, N. (eds) *Nutritional Qualities of Fresh Fruits and Vegetables*. Futura Publications, Mt Kisco, New York, pp. 87–109.

Stewart, J.K. and Aharoni, Y. (1983) Vacuum fumigation with ethyl formate to control the green peach aphid in packaged head lettuce. *Journal of the American Society for Horticultural Science* 108, 295–298.

Stone, W.J.H. and Nelson, M.R. (1966) Alfalfa mosaic virus (calico) of lettuce. *Plant Disease Reporter* 50, 629–631.

Stout, A.B. (1916) Self- and cross-pollination in *Cichorium intybus* with reference to sterility. *Memoir of the New York Botanical Garden* 6, 333–454.

Stout, A.B. (1917) Fertility in *Cichorium intybus*: the sporadic occurrence of self-fertile plants among the progeny of self-sterile plants. *American Journal of Botany* 4, 375–395.

Stubbs, L.L. and Grogan, R.G. (1963) Necrotic yellows: a newly recognized virus disease of lettuce. *Australian Journal of Agricultural Research* 14, 439–459.

Subbarao, K.V., Koike, S.T. and Hubbard, J.C. (1996) Effects of deep plowing on the distribution and density of *Sclerotinia minor* sclerotia and lettuce drop incidence. *Plant Disease* 80, 28–33.

Subbarao, K.V., Hubbard, J.C., Greathead, A. and Spencer, G.A. (1997) Verticillium wilt. In: Davis, R.M., Subbarao, K.V., Raid, R.N. and Kurtz, E.A. (eds) *Compendium of Lettuce Diseases*. American Phytopathological Society, St Paul, pp. 26–27.

Suzuki, Y. (1981) After-ripening as a factor in lettuce seed germination response. *American Journal of Botany* 68, 859–863.

Takeba, G. (1983) Rapid decrease in the glutamine synthetase activity during imbibition of thermodormant New York lettuce seeds. *Plant and Cell Physiology* 24, 1469–1476.

Tesi, R. (1968) Miglioramento genetico della scarola (*Cichorium endivia* L. v. *latifolium* Hegi). *Rivista di Ortoflorofrutticoltura* 4, 416–444.

Thibodeau, P.O. and Minotti, P.L. (1969) The influence of calcium on the development of lettuce tipburn. *Journal of the American Society for Horticultural Science* 94, 372–376.

Thompson, R.C,. (1926) *Tipburn of Lettuce.* Bulletin 311, Colorado Experiment Station, Fort Collins, Colorado, USA, 31 pp.

Thompson, R.C. (1938) *Genetic Relations of Some Color Factors in Lettuce.* Technical Bulletin No. 620, US Department of Agriculture, Washington, DC, 38 pp.

Thompson, R.C. (1946) Germination of endive seed (*Cichorium endivia*) at high temperature stimulated by thiourea and by water treatments. *Proceedings of the American Society for Horticultural Science* 47, 323–326.

Thompson, R.C. and Horn, N.L. (1944) Germination of lettuce seed at high temperature (25 to 35°C) stimulated by thiourea. *Proceedings of the American Society of Horticultural Science* 45, 431–439.

Thompson, R.C. and Ryder, E.J. (1961) *Descriptions and Pedigrees of Nine Varieties of Lettuce.* Technical Bulletin No. 1244, Agricultural Research Service, US Department of Agriculture, Washington, DC, 19 pp.

Thompson, R.C., Whitaker, T.W. and Kosar, W.F. (1941) Interspecific genetic relationships in *Lactuca. Journal of Agricultural Research* 63, 91–107.

Thompson, R.C., Whitaker, T.W., Bohn, G.W. and Van Horn, C.W. (1958) Natural cross-pollination in lettuce. *Proceedings of the American Society for Horticultural Science* 72, 403–409.

Thompson, T.L. and Doerge, T.A. (1996) Nitrogen and water interactions in subsurface trickle-irrigated leaf lettuce II. Agronomic, economic, and environmental outcomes. *Soil Science Society of America Journal* 60, 168–173.

Tibbitts, T.W., Morgan, D.C. and Warrington, I.J. (1983) Growth of lettuce, spinach, mustard, and wheat plants under four combinations of high-pressure sodium, metal halide, and tungsten halogen lamps at equal PPFD. *Journal of the American Society for Horticultural Science* 108, 622–630.

Tomlinson, J.A. and Faithfull, E.M. (1979) Effects of fungicides and surfactants on the zoosopores of *Olpidium brassicae. Annals of Applied Biology* 93, 13–19.

Tomlinson, J.A. and Garrett, R.G. (1964) Studies on the lettuce big-vein virus and its vector *Olpidium brassicae* (Wor.) Dang. *Annals of Applied Biology* 54, 45–61.

Townsend, G.R. (1934) *Bottom Rot of Lettuce.* Memoir 158, Cornell University Agricultural Experiment Station, Ithaca, New York, USA, 46 pp.

Tracy, W.W., Jr (1904) *American Varieties of Lettuce.* Bulletin No. 69. US Department of Agriculture, Washington, DC, USA, 103 pp.

Ulbright, C.E., Pickard, B.G. and Varner, J.E. (1982) Effects of short chain fatty acids on radicle emergence and root growth in lettuce. *Plant, Cell, and Environment* 5, 293–301.

USDA (1998) *Vegetables 1997 Summary.* National Agricultural Statistics Service, Washington, DC, USA.

Valdes, V.M. and Bradford, K.J. (1987) Effects of seed coating and osmotic priming on the germination of lettuce seeds. *Journal of the American Society for Horticultural Science* 112, 153–156.

Valette, R. (1978) Influence de la température sur la germination des semences de chicorée de Bruxelles. *Bulletin des Recherches Agronomique Gembloux* 13, 183–196.

van Bruggen, A.H.C., Brown, P.R. and Greathead, A. (1990a) Distinction between infectious and noninfectious corky root of lettuce in relation to nitrogen fertilizer. *Journal of the American Society for Horticultural Science* 115, 762–770.

van Bruggen, A.H.C., Brown, P.R., Shennan, C. and Greathead, A.S. (1990b) The effect of cover crops and fertilizaton with ammonium nitrate on corky root of lettuce. *Plant Disease* 74, 584–589.

van Bruggen, A.H.C., Jochimsen, K.N. and Brown, P.R. (1990c) *Rhizomonas suberifaciens* gen. nov., sp. nov., the causal agent of corky root of lettuce. *International Journal of Systematic Bacteriology* 40, 175–188.

Vanderwoude, W.J. and Toole, V.K. (1980) Studies of the mechanism of enhancement of phytochrome-dependent lettuce seed germination by prechilling. *Plant Physiology* 66, 220–224.

Van Holsteijn, H.M.C. (1980) Growth of lettuce. II. Quantitative analysis of growth. *Mededeling Landbouwhogeschool, Wageningen* 80, 1–24.

Van Staden, J. (1973) Changes in endogenous cytokinins of lettuce seed during germination. *Physiologia Plantarum* 28, 222–227.

Varoquaux, P., Mazollier, J. and Albagnac, G. (1996) The influence of raw material characteristics on the storage life of fresh-cut butterhead lettuce. *Post Harvest Biology and Technology* 9, 127–139.

Varotto, S., Pizzoli, L., Lucchin, M. and Parrini, P. (1995) The incompatibility system in Italian red chicory (*Cichorium intybus* L.). *Plant Breeding* 114, 535–538.

Verkerk, K. and Spitters, C.J.T. (1973) Effects of light and temperature on lettuce seedlings. *Netherlands Journal of Agricultural Science* 21, 102–109.

Vermeulen, A., Desprez, B., Lancelin, D. and Bannerot, H. (1994) Relationships among *Cichorium* species and related genera as determined by analysis of mitochondrial RFLPs. *Theoretical and Applied Genetics* 88, 159–166.

Vertucci, C.W. and Roos, E.E. (1990) Theoretical basis of protocols for seed storage. *Plant Physiology* 94, 1019–1023.

Vetten, H.J., Lesemann, D.E. and Dalchow, J. (1987) Electron microscopical and serological detecton of virus-like particles associated with lettuce big vein disease. *Journal of Phytopathology* 120, 53–59.

Villiers, T.A. and Edgcumbe, D.J. (1975) On the cause of seed deterioration in dry storage. *Seed Science and Technology* 3, 761–774.

Von der Pahlen, A. and Crnko, J. (1965) El virus del mosaico de la lechuga (*Marmor lactucae*, Holmes) en Mendoza y Buenos Aires. *Revista de la Investigaciones Agropecuarias, Buenos Aires* 2, 25–31.

Vovlas, C. and Quacquarelli, A. (1975) Le virosi delle piante ortensi in Puglia. XVII. Um giallume della Lattuga indotto da virus della maculatura gialla Cicoria. *Phytopathologia Mediterranea* 14, 144–146.

Vovlas, C., Martelli, G.P. and Quacquarelli, A. (1971) Le virosi delle piante ortensi in Puglia. VI. Il complesso delle maculature anulari della Cicoria. *Phytopathologia Mediterranea* 10, 244–254.

Walkey, D.G.A. (1967) Chlorotic stunt of lettuce caused by arabis mosaic virus. *Plant Pathology* 16, 20–23.

Walkey, D.G.A. and Pink, D.A.C. (1990) Studies on resistance to beet western yellows virus in lettuce (*Lactuca sativa*) and the occurrence of field sources of the virus. *Plant Pathology* 39, 141–155.

Walsh, J.A. (1994) Effects of some biotic and abiotic factors on symptom expression of lettuce big-vein virus in lettuce (*Lactuca sativa*). *Journal of Horticultural Science* 69, 21–28.

Wang, M., Cho, J.J., Provvidenti, R. and Hu, J.S. (1992) Identification of resistance to tomato spotted wilt virus in lettuce. *Plant Disease* 76, 642.

Watts, L.E. (1975) The response of varous breeding lines of lettuce to beet western yellows virus. *Annals of Applied Biology* 81, 393–397.

Waycott, W. and Taiz, L. (1991) Phenotypic characterization of lettuce dwarf mutants and their response to applied gibberellins. *Plant Physiology* 95, 1162–1168.

Waycott, W., Fort, S.B. and Ryder, E.J. (1995) Inheritance of dwarfing genes in *Lactuca sativa* L. *Journal of Heredity* 86, 39–44.

Weber, G.F. and Foster, A.C. (1928) *Diseases of Lettuce, Romaine, Escarole, and Endive.* Bulletin 195, University of Florida Agricultural Experiment Station, Gainsville, Florida, USA, pp. 303–333.

Welch, J.E., Grogan, R.G., Zink, F.W., Kihara, G.M. and Kimble, K.A. (1965) Calmar, a new lettuce variety resistant to downy mildew. *California Agriculture* 19, 3–4.

Westerlund, F.V., Campbell, R.N. and Grogan, R.G. (1978a) Effect of temperature on transmission, translocation, and persistence of the lettuce big-vein agent and big-vein symptom expression. *Phytopathology* 68, 921–926.

Westerlund, F.V., Campbell, R.N., Grogan, R.G. and Duniway, J.M. (1978b) Soil factors affecting the reproduction and survival of *Olpidium brassicae* and its transmission of big-vein agent to lettuce. *Phytopathology* 68, 927–935.

Wheeler, T.R., Hadley, P., Morison, J.I.L. and Ellis, R.H. (1993) Effects of temperature on the growth of lettuce (*Lactuca sativa* L.) and the implications of assessing the impacts of potential climate change. *European Journal of Agronomy* 2, 305–311.

Whitaker, T.W. and Jagger, I.C. (1939) Cytogenetic observations in *Lactuca. Journal of Agricultural Research* 58, 297–306.

Whitaker, T.W. and Pryor, D.E. (1941) The inheritance of resistance to powdery mildew (*Erysiphae cichoracearum*) in lettuce. *Phytopathology* 31, 534–540.

Whitaker, T.W., Kishaba, A.N. and Toba, H.H. (1974a) Host-parasite interrelations of *Lactuca saligna* L. and the cabbage looper, *Trichoplusia ni* (Hubner). *Journal of the American Society for Horticultural Science* 99, 74–78.

Whitaker, T.W., Ryder, E.J., Rubatzky, V.E. and Vail, P.V. (1974b) *Lettuce Production in the United States.* Agriculture Handbook 221, United States Department of Agriculture, Agricultural Research Service, Washington, DC, USA, 443 pp.

Wiebe, H.J. (1989) Effects of low temperature during seed development on the mother plant on bolting of vegetable crops. *Acta Horticulturae* 253, 25–30.

Willocx, F., Hendrickx, M. and Tobback, P. (1993) Evaluation of safety and quality of 'Grade 4' products. In: *Proceedings of the Workshop Cost 95, Leuven, Belgium, 14–15 September 1993*, pp. 195–213.

Wong, T.K., Harper, F.C. and Mai, W.F. (1970) Soil fumigation for controlling root knot of lettuce on organic soils. *Plant Disease Reporter* 54, 368–370.

Wurr, D.C.E. and Fellows, J.R. (1984) The effects of grading and 'priming' seeds of crisphead lettuce cv. Saladin, on germination at high temperature, seed 'vigour' and crop uniformity. *Annals of Applied Biology* 105, 345–352.

Wurr, D.C.E. and Fellows, J.R. (1987) The laboratory germination of crisp lettuce seeds under moisutre stress. *Annals of Applid Biology* 111, 445–453.

Wurr, D.C.E. and Fellows, J.R. (1991) The influence of solar radiation and temperature on the head weight of crisp lettuce. *Journal of Horticultural Science* 66, 183–190.

Wurr, D.C.E., Fellows, J.R. and Drew, R.L.K. (1987a) The germination and the forces required to penetrate seed layers of different seedlots of three cultivars of crisp lettuce. *Annals of Applied Biology* 110, 405–411.

Wurr, D.C.E., Fellows, J.R. and Pittam, A.J. (1987b) The influence of plant raising conditions and transplant age on the growth and development of crisp lettuce. *Journal of Agricultural Science, Cambridge* 109, 573–581.

Wurr, D.C.E., Fellows, J.R. and Suckling, R.F. (1988) Crop continuity and prediction of maturity in the crisp lettuce variety Saladin. *Journal of Agricultural Science, Cambridge* 111, 481–486.

Wurr, D.C.E., Fellows, J.R., Hiron, R.W.P., Antill, D.N. and Hand, D.J. (1992) The development and evaluation of techniques to predict when to harvest iceberg lettuce heads. *Journal of Horticultural Science* 67, 385–393.

Zalom, F.G. (1981) The influence of reflective mulches and lettuce types on the incidence of aster yellows and abundance of its vector, *Macrosteles fascifrons* (Homoptera: Cicadellidae), in Minnesota. *The Great Lakes Entomologist* 14, 145–150.

Zerbini, F.M., Koike, S.T. and Gilbertson, R.L. (1997) *Gazania* spp.: a new host of lettuce mosaic potyvirus, and a potential inoculum source for recent lettuce mosaic outbreaks in the Salinas Valley of California. *Plant Disease* 81, 641–646.

Zink, F.W. (1955) Studies with pelleted lettuce seed. *Proceedings of the American Society for Horticultural Science* 65, 335–341.

Zink, F.W. (1967) Effect of bed direction on growth and harvest density of head lettuce. *Proceedings of the American Society for Horticultural Science* 91, 369–376.

Zink, F.W. and Duffus, J.E. (1969) Relationship of turnip mosaic susceptibility and downy mildew (*Bremia lactucae*) resistance in lettuce. *Journal of the American Society for Horticultural Science* 94, 403–407.

Zink, F.W. and Duffus, J.E. (1970) Linkage of turnip mosaic virus susceptibility and downy mildew, *Bremia lactucae*, resistance in lettuce. *Journal of the American Society for Horticultural Science* 95, 420–422.

Zink, F.W. and Kimble, K.A. (1960) Effect of time of infection by lettuce mosaic virus on rate of growth and yield in Great Lakes lettuce. *Proceedings of the American Society for Horticultural Science* 76, 448–454.

Zink, F.W. and Yamaguchi, M. (1962) Studies on the growth rate and nutrient absorption of head lettuce. *Hilgardia* 32, 471–500.

Zink, F.W., Grogan, R.G. and Welch, J.E. (1956) The effect of percentage of seed transmission upon subsequent spread of lettuce mosaic virus. *Phytopathology* 46, 662–664.

Zitter, T.A. and Guzman, V.L. (1974) Incidence of lettuce mosaic and bidens mottle virus in lettuce and escarole fields in Florida. *Plant Disease Reporter* 58, 1087–1091.

Zitter, T.A. and Guzman, V.L. (1977) Evaluation of cos lettuce crosses, endive cultivars, and *Cichorium* introductions for resistance to bidens mottle virus. *Plant Disease Reporter* 61, 767–770.

Zohary, D. (1991) The wild genetic resources of cultivated lettuce (*Lactuca sativa* L.). *Euphytica* 53, 31–35.

INDEX

abscisic acid 60
acetaldehyde 151
achene 9–11, 50, 54, 67
action spectra 56
adaptation 41, 43, 45–46, 50, 79,
    89, 135
Agrobacterium 32, 46
    A. tumefaciens 35
alginate 114
α-naphthaleneacetic acid 62
Alternaria 37
    A. cichorii 115–116
Amaranthus spinosus 160
amphidiploids 13
ancymidol 56
animal fodder 26
anthesis 68, 110
anthocyanin 8, 11, 18, 21, 31–32,
    41, 166
anthracnose 42, 114, 125, 147
arabis mosaic 114, 144
asparagus lettuce 17, 22
aster leafhopper 155
Asteraceae 8, 12, 56, 70, 147, 153,
    159–161
auction 170
autohydrolysis 55

backcrossing 38, 40
bacterial soft rot 75, 77–78
Barbarea vulgaris 143
barbe de capucin 25
Batavia(n) lettuce 5, 18–20, 30, 46,
    63, 139–140, 145, 167
beet armyworm 156–157
beet pseudo–yellows 139, 141, 154
beet western yellows 139–140,
    144, 147, 150–151
beet yellow stunt 139–140
Belgian endive 26, 50, 172
Bemisia argentifolii 140, 155
Bemisia tabaci 140, 154–155
benzyladenine 60
Bibb subtype 21–22
bidens mottle 30, 47, 144, 147
Bidens pilosa 144
biennial 11, 70, 115–116
big vein 42–43, 136–138, 147
Blapstinus spp. 158
bolting 20, 33, 42, 49, 50, 67,
    71–72, 89–90, 110, 115
Boston subtype 21
botrytis 42–43, 78
Botrytis cinerea 77
brand label 101